Attack Evangelism

Attack Evangelism

Being on the offense without being offensive

Greg Koehn Ph.D.

AuthorHouse™ LLC
1663 Liberty Drive
Bloomington, IN 47403
www.authorhouse.com
Phone: 1-800-839-8640

Published by AuthorHouse   03/28/2014

ISBN: 978-1-4918-6965-9 (sc)
ISBN: 978-1-4918-6964-2 (e)

Library of Congress Control Number: 2014904134

TABLE OF CONTENTS

INTRODUCTION

Most books written on evangelism follow a more passive approach. It is about sharing Jesus, live the lifestyle and people will just come to you and ask, and it is about being able to do it in such a low key, easy manner that you do not need to be afraid of offending anyone. One of my mentors was an evangelist and he was asked if he was ever afraid of driving people away. He replied, where are we going to drive them, they are already going to hell? There are unintended consequences to this passive approach. First, it is so passive that people do not get around to actually giving the gospel. Second, it is so passive that it is easily ignored and considered to be irrelevant by the masses of people. This great salvation has been reduced down to a fire escape with very little, if any, relevance to daily living.

There are basically two styles of evangelism. The first is known as *life style* evangelism. This style of evangelism has been defined by the author of *Lifestyle Evangelism*, Dr. Joe Aldrich in the following manner, "It is the constant and spontaneous outflow of our individual and corporate experience of Christ. Even more specifically, evangelism is what Christ does through the activity of His children as they are involved in (1) proclamation (2) fellowship, and (3) service"[1] The essence of this style is just as its name suggests, it is living out the life style of a Christian, being ethical in our dealings with others, being sympathetic and empathetic, and then have something to say when the opportunity presents itself. A person is to be a Christian before broadcasting to others about how they should be one as well. The advantage to this approach is that it is non-threatening and is a natural part of our every day life. The disadvantage is that Christians forget to give the gospel and when they do it is so non-threatening that they do not give people the opportunity to trust Christ. This was certainly not the intent of Aldrich or of any other writer of books on evangelism, but it has become the outcome.

The second style of evangelism is *confrontational* evangelism. This style is in your face, opposing, and challenging. The advantage of this style is that it places evangelism as a top priority, it is more aggressive and tends to get more results. The disadvantage is that it tends to offend people and cause people who are already against the Gospel to be more so after the encounter.

I am proposing a third style of evangelism which I call *attack evangelism*. This is an evangelistic mental attitude that is on the offense without being offensive. It is a mind-set

---

[1]   Joe Aldrich, *Life Style Evangelism* (Colorado Springs: Multnomah Books, 1993) 29.

that will recognize and take advantage of opportunities and will create opportunities where none had existed. This is a style that combines the best of both lifestyle and confrontational evangelism. This means that we will pursue opportunities that come as a result of our lifestyle and will create opportunities through confrontation. It is important to keep in mind that this is being on the offense without being offensive. Later in this volume you will be able to see how this works by using questions.

The question might be raised as to why use the word *attack*, that sounds so aggressive and unloving? Should we not be passive? Christians tend to like *retreats*, therefore the idea of an *attack* is a foreign concept. There are at least four reasons why it is important that the Lord's people go on the attack. First, passive evangelism has not worked. For the first time in its history the United States is not a Christian country. Postmodernism is the dominating philosophy of the day. Atheist and homosexuals have both come out of the closet and are spewing their venomous lies without anyone to challenge their presuppositions. Second, we are at war and our warfare is in the spiritual realm, "For we do not wrestle against flesh and blood, but against principalities, against powers, against the rulers of the darkness of this age, against spiritual hosts of wickedness in the heavenly places." (Ephesians 6:12). We see that there are two sides with different world views containing a totally different basis for answering the key questions of life; what is real, what is true, what is the meaning of life and what happens when we die. At the root of the discussion is a spiritual problem which only has a spiritual solution.

Third, Christians have taken on the mantle of judges to point out what is wrong and have pursued political processes to correct the problems. All this has accomplished is to turn the Church into a political movement with motives that are suspect. Politics will always divide and one side must win while the other side loses. It is time for Christians to make a stand for what they are for, that faith in Christ not only provides eternal life but a better and more prosperous life in the here and now. Fourth, those who do not place their faith in Christ will go to hell. This is reality and hell is a very real place. How can anyone know that there is a better way with a better destination unless someone tells them (Romans 10:14-15)? To attack with the gospel is actually the greatest gift of love that can be given.

Due to the high regard of the author for the New Testament Church the word *Church* will invariably be capitalized. This volume is submitted to give a theological and philosophical basis for giving the gift of the gospel. May we proceed with the understanding that the only thing that we cannot do in heaven is give the gospel. Therefore, it is time to attack!

CHAPTER ONE

THIS GREAT SALVATION

Once while going door to door in North Omaha, I came across a man who was sitting in a car drinking out of a bottle in a brown paper bag. Hoping that the man was not too far gone due to the intoxication of the contents of his bottle, I pressed forward. I asked him if anyone had ever taken a Bible and shown him how he could know for sure that he is going to heaven, May I? He replied that first he would like to know the answer to a certain question. His question was, is it true that black people were created on Friday 13, which would explain why they are so unlucky? Obviously the man had been exposed to bad theology and was in need of instruction. It is good for us to understand the foundational truths of what we believe because you never know when you may be in need of sound theology to overcome poor theology. It is also a bit risky, because our theology will demand that we preach the one way to the one God.

Our Soteriology (the study of our salvation) can be summarized in I Timothy 2:5-6 "For there is one God and one Mediator between God and men, the Man Christ Jesus, who gave Himself a ransom for all, to be testified in due time,". This verse can be broken down into three components i.e. there is **one God**; there is **one mediator**; and there is **one humanity**. Therefore, evangelism is about how humanity is separated from God and how God has made it possible for man to be re-connected to God by the one way that God has determined. Our message of good news, is to tell how there is one way to the one God and that has to be God's prescribed way.

This message begins with the very first verse of the Bible. In Genesis 1:1 we read, "In the beginning God . . .". The Bible begins with the assumption that God exists, there is nothing here that seeks to explain or defend the existence of God. As the text moves forward we are given what is known as the cosmological argument for the existence of God, an argument we will explain in a later chapter, but at the beginning it is assumed that there is a God and that He created the universe.

This brings us to the question of **why bother**? Why does God bother to be sure that all of this is written down and preserved over the period of thousands of years? There are those, known as deists who believe that there is a god who is a creator but who has not been involved in his creation since. In other words he is like a watch maker who makes the watch, winds it up and then forgets about it. This is not true of God as He has revealed Himself in His creation and in His Word. In His Word He has shown how that He is very much involved in creation and the events of humanity. He has also revealed how mankind became estranged from Him and how there is reconciliation through the sacrifice of His son.

As the Scriptures unfold we have further insight into the **one God**. In the book of Exodus Moses is seen on the backside of the desert. He sees a bush that will not burn up in contrast to what was usually the case with Acacia bushes that would explode into flames and then be gone. He went to see this amazing sight and God spoke to him from the bush. God explained to Moses that he was to go to Pharaoh and demand the release of Israel from their bondage. Moses then asked, when the people of Israel ask what is the name of God, what should he tell them? The answer is most instructive, "**I AM WHO I AM**." A common response to that might be, " I am what?". God will go on to explain to Moses that He is the God of Abraham, Isaac and Jacob and that should suffice. But Israel has been in captivity for over 400 years and they have been surrounded by the many false gods of Egypt, therefore they are in need of this clarification. However, it is important to note this phrase, "I AM WHO I AM" as it indicates to us that **He is self-contained and all inclusive.** There is nothing that can be added to what He is, He is in need of nothing to make Him more complete, and He is perfect in every way. This being the case, everything He does, He does according to His sovereign will, not because He must do something to satisfy a lack in His person. To put it more simply, He does what He does because He wants to not because He has to.

We see that God works within the framework of His sovereign will in Exodus 4:21, "But I will harden his heart, so that he will not let the people go." First Moses was told to go to Pharaoh and by signs and wonders tell him to let Israel go but at the same time God would harden his heart so that he will not let them go. In the book of Romans we have a commentary on this very text where we read concerning the hardening of Pharaoh's heart, "Therefore He has mercy on whom He wills, and whom He wills He hardens." (Romans 9:18). At this point of our learning we simply want to keep in mind that salvation is of the Lord, it is according to His divine plan and will. We will explore this thought more as we go along.

We are told more about the nature of God in the giving of the law, "For I, the Lord your God, am a jealous God, visiting the iniquity of the fathers upon the children to the third and fourth generations of those who hate me." (Deuteronomy 5:9). There are to be no other gods before

Him in our worship and affections because there are no other gods before Him. There is only one God and He alone is to be worshipped, obeyed and loved Deuteronomy 6:5. There is nothing that is His equal Isaiah 40:21-26.

Having established that there is only one God, it is now important to understand that there is only **one way to that one God**. Going back to the book of Exodus we have a great illustration of the one way to the one God in the tabernacle of the wilderness. It is worthy of note that the instructions concerning the tabernacle and the priesthood take up 2/3 of the entire book. Then it is important to notice that the tabernacle was a copy and shadow of the things in heaven Hebrews 8:5. The over-riding lesson of the tabernacle is to teach how a man can get into the presence of God. The instructions of the tabernacle were given from the inside out as God speaks from Himself out, but for mankind, we must work from the outside in. Therefore the first thing we notice is that there is a wall of white linen surrounding the tabernacle. This wall of white linen represents perfection, so the way to God is blocked by perfection until you get to the gate. Here is the way in and the first thing that confronts a person is the brass altar; followed by the laver; then the door to the holy place with the lamp stand, table of showbread and the altar of Incense; then the veil to the holy of holies and the ark of the covenant. Since it is not within the purpose of this book to do a study on the tabernacle, it will suffice to say that the tabernacle serves as an excellent picture of the **one way to the one God as defined by God!**

Before leaving the picture of the tabernacle there are two more points worthy of consideration. First, it is important to keep in mind that under the law, access to God was restricted to the High Priest, once per year on the day of atonement Leviticus 16, "Tell Aaron your brother not to come at just any time into the Holy Place inside the veil, before the mercy seat which is on the ark, lest he die; for I will appear in the cloud above the mercy seat." (vs. 2). This stands in contrast to we who have access to the Father at any time Romans 5:1-2; by way of a priest who ever lives to make intercession on our behalf Hebrews 7:25-27.

The second point of importance is that access to God is by way of sacrifice and the shedding of blood. From the beginning of our Bible we pick up on a trail of blood. Abel offered a sacrifice from his flock with its fat which is a bloody sacrifice and God had respect to that offering but Cain offered a sacrifice from the fields which had no blood and that sacrifice was rejected Genesis 4:3-8. Within the Noahic covenant the significance of blood is clearly defined, "But you shall not eat flesh with its life, that is, its blood." the blood is the life and "Whoever sheds man's blood, by man his blood shall be shed; For in the image of God He made man." (Genesis 9:4, 6). This is where capital punishment was instituted that whoever kills another i.e. sheds their blood, then their blood must be shed. The wages of sin is death and the sign that the wage has been paid is by the shedding of blood. We have this principle again in Exodus 12 with the Passover where we read, "And when I see the blood, I will pass over you;" (vs. 13) the redemption or purchase price for the first born in Egypt was the blood of a sacrificial lamb. A life was to be given up, either the first born or the lamb as the substitute.

The definition that the life is in the blood is expanded upon in Leviticus 17:11 "For the life of the flesh is in the blood, and I have given it to you upon the altar to make atonement for your souls; for it is the blood that makes atonement for the soul." In this text we have the added feature that God has provided an altar with the blood to make atonement. In the Old Testament sacrificial system we see the blood of the sacrifice is sprinkled everywhere, "And according to the law almost all things are purified with blood, and without shedding of blood there is no remission." (Hebrews 9:22). The law with the tabernacle, then the temple and all the sacrifices included in the Levitical system could only cover sin for a season. The law along with these sacrifices pointed forward to a better priesthood based upon a better sacrifice. It is the sacrifice of our Lord Jesus that is the better sacrifice that establishes the better priesthood. This will be examined more thoroughly in the section of the one mediator.

The sum of what we have so far is that there is only one way to the one God and that way has been prescribed by God to be the way of sacrifice and the shedding of blood. There is no other way to God, anyone who comes to God must come to Him by way of His Son who stated, "I am the way, the truth, and the life. No one comes to the Father except through Me." (John 14:6). It is important to understand how emphatic is this statement. There is a definite article in front of each word, way, truth, and life. This makes it emphatic that there is no other way to God but through the Lord. We dare not even imply that there is more than one way to God.

This brings us to the **one mediator** between God and man, the man Christ Jesus. In the story of Job we have his complaint in 9:33, "Nor is there any mediator between us, Who may lay his hand on us both." Sin has separated man from God and Job felt at the time that there was no one to stand in the gap between God and man as the mediator, the one who could/would lay his hand upon them both. But later in the story Job receives better insight when in chapter 19 he declares, "I know that my redeemer lives, and that He shall stand at last on the earth;". He is made to understand that he does have a redeemer who will make it possible for him to see God, "And after my skin is destroyed, this I know, That in my flesh I shall see God, Whom I shall see for myself, And my eyes shall behold, and not another." (vs. 26,27). It is our Lord who is the redeemer, the one who stands in the gap between God and man. He is the one who satisfies the just claims of a holy God and meets the need of all humanity. It is in Him that "Mercy and truth have met together; righteousness and peace have kissed." (Psalms 85:10). This is how He came to earth in His incarnation as the one who is full of Grace and Truth John 1:14.

Later in this chapter we will examine terms common to our salvation which will elaborate more on the person and work of our Lord as the mediator. Again we emphasize that He is the **only mediator between God and man.** He is the only one who is the Word made flesh, fully God and yet fully man. The title of "only begotten" belongs only to Him as He is one of a kind, distinct from all others who would be called the sons of God.

This truth is laid out for us more fully in Colossians 1:15-22. It behooves us now to note some of the highlights of this wonderful text. We will do this by noting key words and phrases which speak of our Lord.

- He is the image of the invisible God. He is the exact reproduction of God in a physical form i.e.

the Word become flesh; He is the manifestation of the hidden or unseen. God is seeking to reveal Himself in terms that man can understand which is His Son.

- He is the firstborn over all creation. He is not only prior to all creation, He is sovereign over all creation.

- For by Him all things were made. The word *by* is translated from the word *en* (en) which is not instrumental but locative. As Vincent puts it, "*In* is not *instrumental* but *local*; not denying the instrumentality, but putting the fact of creation with reference to its sphere and center. *In Him*, within the sphere of His personality, resides the Creative will and the creative energy, and in that sphere the creative act takes place. Thus creation is dependent on Him."[2] Therefore, we understand that creation came out of Him, He is the origin of all things that exist.

- All things were created through Him and for Him. He is the intermediate agency of all things and all things were created to find their consummation in Him. Therefore, all things begin and end with Him.

---

2   Kenneth S. Wuest, *Wuest's Word Studies Vol.1* (Grand Rapids: Eerdmans Publishing Co. 1953), 183.

- He is before all things. This states His pre-existence. He is the unmoved mover, the uncaused cause, before all else He was already in existence without a beginning of His own.

- In Him all things consist. He not only created all things, He holds all things together. As in Hebrews 1:3, ". . . upholding all things by the word of His power . . .". He is the creator and the sustainer of all things.

- He is the head of the Church. He is not only over all things physical but all things spiritual as well. He is the head of the church as the preeminent one and no man has the right to lay claim to that title in way shape or form.

Therefore, He is uniquely qualified to be the one who stands between God and man. He has reconciled us to God. That fellowship which existed between God and humanity at creation was lost through sin but is now restored through the finished work of our Lord, "And you, who once were alienated and enemies in your mind by wicked works, yet now He has reconciled in the body of His flesh through death, to present you holy, and blameless, and above reproach in His sight." (Colossians 1:21, 22). In verse 20 of the same text, we see how that all things have been reconciled to Him and the result is peace. We read in Romans 8 that all creation is subject to futility and is groaning until such time as our salvation has come to its full measure. So then this reconciliation is for us who have believed and will have implications for all creation. This reconciliation will bring peace, i.e. the bringing together of two estranged parties with the result of a cessation of hostilities. It is worth noting that it is we who were alienated and enemies who have been reconciled. The word *"alienated"* is passive in voice which denotes the state or position and the word *"enemies"* is active in voice which denotes the activity. Therefore what we have here is that in our past state we were in a position of defiance against God and we were actively engaged in those things that were against God. You cannot get worse. But we have been reconciled in the person and work of our Lord.

It has been previously described how reconciliation can only come about by the shedding of blood. Therefore, an essential aspect to the work of our Lord Jesus is in the shedding of His blood. The book of Hebrews gives a very clear argument as to how there is only one sacrifice for sin and if one misses that one sacrifice then there is no other sacrifice available. Beginning in chapter two and verse 3, ". . . how shall we escape if we neglect so great a salvation, which at the first began to be spoken by the Lord, and was confirmed to us by those who heard Him." The point being is that there is no escape because there is no other salvation. In Hebrews 6 and verses 4 to 6 we have verses that are many times terribly misunderstood and therefore mistakenly applied. "For it is impossible for those who are once enlightened, and have tasted the heavenly gift, and have become partakers of the Holy Spirit, and have tasted the good word of God and the powers of the age to come, if they fall away, to renew them again to repentance; since they crucify again for themselves the Son of God, and put Him to an open shame." The very first thing that is important to understand in this text are the words "For it is impossible . . . if they fall away, to renew them again to repentance,". In between these two statements there are parenthetical statements describing this person. The theme of Hebrews is Christ is better than the Old Testament law and covenant. In keeping with this theme the argument is made that the sacrifice of Christ is perfect, therefore there is nothing that can be added to it, you cannot make something more perfect. Therefore, hypothetically speaking, if it were possible for a person to fall away they could never be renewed to repentance because there is no other sacrifice for sins. The sacrifice of our Lord is once and for all and faith in that one sacrifice is once and for all. This simply means that if it were possible for a person to lose their salvation it would not be possible for them to get it back. Fortunately we know from many other Scriptures that it is not possible to lose this great salvation.

In chapters nine and ten of Hebrews there is a wonderful discourse on the value of the blood of Christ. This is important in understanding His work as the one mediator between God and man. In chapter nine and verses 11 to 15 Christ is the High Priest of a greater and more perfect tabernacle, meaning that His priesthood is in heaven and is spiritual. The Old Testament priesthood was of this earth and physical. The Old Testament priesthood used the blood of animals for redemption and this redemption was not permanent, it needed to be repeated over and over again. But the blood of Christ is the better sacrifice because it was made by the Son of God through the eternal Spirit of God, offered to God the father. Just as the entire trinity was at work in creation so also the entire trinity is at work in the atonement. It is through the blood of the Lord Jesus that our conscience is cleansed from dead works to serve the living God. It is for this reason that He is the mediator of the new covenant (vs 15). The old covenant was established by the blood of animals but the new covenant is established by the blood of Christ. This means that the new covenant is superior because it is established by a superior sacrifice. The principle that all things are purified with the shedding blood (vs 23) is kept in force with the blood of Christ. His sacrifice and His blood is so far superior that all the things that make up the Old Testament worship are nothing more than a copy of the true which is in heaven (vs 24).

In chapter ten and verses 5-18 we read of how when He came into the world, He declared that it was to do the will of God. This is what I call the Christmas story from the perspective of heaven. The little baby in the manger declared that He had come to do the will of God. Think what a shock that must have been to Mary. This is probably not what happened but nevertheless Mary knew that this was no ordinary baby. The body that was prepared for Him was for the purpose of being the sacrifice for sins and after that one sacrifice was made, then He forever sat down at the right hand of the Father. We are **forever** perfected based on that one sacrifice.

Within this context there are at least two principles that are emphasized. The first is that our great salvation was accomplished in accordance with the will of God. As we read in verse 7, ". . . In the volume of the book it is written of me—to do Your will, O God." And verse 10, "By that will we have been sanctified through the offering of the body of the body of Jesus Christ once for all." There is a lot of sentimentality about how we nailed Him to the cross, because it was our sin for which He died. This is true, but the first priority was that it was the will of God. If it were not for that there would have been no sacrifice for sin and all of mankind would go to hell. The second principle is that this one sacrifice has forever perfected those who are sanctified by the will of God. For example we read in verse 12, "But this Man, after He had offered one sacrifice for sins **forever**, sat down at the right hand of God." And in verse 14, "For by one offering He has perfected **forever** those who are being sanctified." This one sacrifice forever sanctifies those who place their trust in Him. There is no other sacrifice and there is no other mediator between God and man.

He is that one mediator, the one way to the one God. This brings us to the **one humanity**. Our anthropology (study of man) begins in the book of Genesis and chapter one. On the sixth day God made man in His own image (Vs 26-31). It is difficult to determine what is meant by this but there are certain conclusions that can be made. Inasmuch as God is a spirit there can be no physical likeness, but at the same time when God appears to man it is as a man esp. the incarnation. We also know that when man was first created he was free from disease, was in perfect health and did not die. Being created in the image of God man had reason, was a moral being and a free agent. Henry Morris describes being in the "image of God" in the following:

> In any case, there can be little doubt that the "image of God" in which man was created must

entail those aspects of human nature which are not shared by animals—attributes such as a moral consciousness, the ability to think abstractly, an understanding of beauty and emotion, and, above all, the capacity for worshiping and loving God. This eternal and divine dimension of man's being must be the essence of what is involved in the likeness of God. And since none of this was a part of the animal *nephesh,* the "soul," it required a new creation.[3]

This phrase, "image of God," represents a great gulf fixed between man and animals. It has been said that you could place one hundred monkeys in front of one hundred type writers for one hundred years and never get a single sentence, much less a whole book. Mankind is far superior to animals in intelligence, ability, morality and emotion. Even more important than that is the ability of man to worship God. You will never see a group of monkeys gathered together to worship God. The Bible speaks to this in Romans 1:18-20,

> For the wrath of God is revealed from heaven against all ungodliness and unrighteousness of men, who suppress the truth in unrighteousness, because what may be known of God is manifest in them, for God has shown it to them, for since the creation of the world His invisible attributes are

---

[3]   Henry M. Morris, *The Genesis Record* (Grand Rapids: Baker Book House, 1976) 74.

clearly seen, being understood by the things that
are made, even His eternal power and Godhead,
so that they are without excuse.

Keil described it in the following manner: "This
consists rather in the fact, that the man endowed with
free self-conscious personality possesses, in his spiritual as
well as corporeal nature, a creaturely copy of the holiness
and blessedness of the divine life."[4]

The word man is generic for both men and women
and applies to the human race throughout the Bible.
In the beginning they were to be fruitful and multiply
and fill up the earth. Together, they were to be stewards
over God's creation to subdue and have dominion. It
appears at the first that they were vegetarians but were
later given permission to eat meat Genesis 9:3. Marriage
is the first social institution and was ordained of God in
2:18-25. Socially man was created with the desire to be in
community and though he was in communion with God
it is written that it was not good for man to be alone and
so a helper was created for him. Here the word *man* speaks
to the man specifically while the word *woman* will speak
to the woman specifically. In the beginning the woman
is to be the one who comes along side of a man to be
his help, his companion, that which completes him and

---

4   C.F.Keil and F. Delitzsch, *Commentary on the Old Testament*,
Vo. 1 (Grand Rapids: Eerdmans Publishing, reprint 1975) 64.

who bears his children. The intelligence of man is seen in that he named the animals, man was not some knuckle dragging descendent of an ape that went around grunting. He and she were upright, intelligent beings superior to all other created creatures and to what men and women are today in every way.

In verses 21-25 we see this very special and unique relationship between the man and the woman. The woman is made from out of the man. Remember that the man is created from the dust of the earth (vs 7) but now God will take a rib out of the man to create the woman. This is where the word *woman* comes from i.e. taken out of the man. When Adam awoke from his deep sleep God presented the woman to him and Adam was very pleased. He announced that this was bone of his bone and flesh of his flesh, a man will now leave his parents and that family unit to form their own family unit. The man and the woman enter into a one flesh union an unbreakable bond. It is important to keep in mind that marriage and the family is the first institution of society. Everything else that frames a society and a culture revolves around the family. When the family structure falls, all of society falls soon thereafter. In creation we constantly see order with everything in its place and a place for everything. The order for the human social construct is a man and a woman joined together as one unbreakable bond who then produce children which forms the family. We also see order within the family. The man was first formed and then the woman from out of the man. The man is in the prominent place. This order of things is stated in I Corinthians 11:3, "But I want you to know that the head of every man is Christ, the head of woman is man, and the head of Christ is God." Note how the head of Christ is God, this is the order of their function and has nothing to do with their equality. Christ is God and is of the same essence as God. Men and women are both made in the image of God and are equal, but in the order of their function, the woman is to submit to the authority of the man (Ephesians 5:22; I Peter 3:1). At the same time the man is to love his wife as Christ loves the Church and respect her as a fellow heir in the grace of life (Ephesians 5:25; I Peter 3:7).

The question might be asked, what does this have to do with our great salvation? The answer is found in Ephesians 5:25-27 where we see marriage as a picture of Christ and the Church. A husband is to love his wife in the same manner as Christ loved the Church and gave Himself for her. The love that our Lord has for us is a self-sacrificing love to where He died for us, was buried and then raised up again and will one day come and take us for His bride throughout all eternity. Therefore, marriage and the family not only are key for our social construct but also as a testimony of this great salvation. But then came the fall.

The account of the fall of man is one of the most well known stories in the Bible and could be one of the most misunderstood or at least underrated. The basic premise of the story is that God commanded that they may eat of all the trees in the garden but of one tree they were not to eat. This tree was the tree of the knowledge of good and evil Genesis 2:16-17 and on the day they ate of it they would surely die. There are a number of items here worth noting. First, they may eat of all the trees in the garden and it can be assumed that there were lots of trees with great variety. It was easy for them to see that God is good and has supplied their every need abundantly above what they could ask or think. They could not eat of just one of the many trees in the garden but nevertheless that is the one which will command all the attention. Second, it is the tree of the knowledge of good and evil. What is so fascinating about evil? They already knew good, why would anyone want to know evil? Third, the command was to not eat of this one tree. There was nothing particular about the fruit of this tree or that it was inherently evil, it was the command to not eat of it. The choice given to Adam and Eve was to obey God in this or disobey. The question is why did God give the choice in the first place? Why not leave well enough alone and have created beings who automatically do the will of God without the potential to do otherwise?

In our chapter on apologetics we will discuss some of these issues more thoroughly but for now we will suffice it to say that for reasons known only to God, He wanted to be loved and making the choice to obey Him will prove that man loves Him. Therefore, man must be free to make a choice that is available to him and God must give man that freedom and the opportunity to make that choice. It is important to keep in mind that God does not need our love because He is self-contained, but He desires our love.

We have the account of the fall then in chapter 3 of Genesis where we have Eve taking a walk in the garden in the vicinity of the forbidden tree and then engage in a conversation with the serpent. Once again it is important to pause for a moment and take note of some salient points. For example, what was Eve doing walking around in the area of the forbidden tree? Where was Adam when all this was going on? Was a talking snake an odd occurrence or did all animals talk before the fall?

Let's take a look at the main characters of our drama and at the events that transpired. The serpent was more cunning than any of the other beast of the field and in the garden of Eden was probably a creature of great beauty. It has been said that there is evidence that at one time the snake could stand upright as some will do on occasion even now. We do not know if Satan took over this particular serpent or if he simply appeared as a serpent as he is capable of changing forms in his appearances. We read of the fall of Satan (Lucifer) in Isaiah 14:12-17 which apparently took place prior to this time in Eden. Satan is called the serpent in Revelation 12:9,14,15; 20:2. Keep in mind that in function Adam was always to be the head and the one responsible for the family. This is the reason why when God came looking for them in the garden, He called for Adam. It is also the reason why even though Eve was first deceived, it is in Adam that all die I Corinthians 15: 22 and it is Adam who is the federal head of all mankind in that through the transgression of Adam all have sinned and all died Romans 5:12-21. I once heard a woman say that she could not believe in a God who would blame a woman for sin in the world. As is usually the case, her unbelief was based upon her ignorance.

The temptation began with the Serpent asking a question which led the woman to doubt both the word of God and the goodness of God. "Has God indeed said, 'You shall not eat of every tree of the garden'?" (2:1). Eve's response was that they could not eat nor touch the forbidden tree or they would surely die. God had said nothing about touching the tree, but she added to what God had said making the requirement even more stringent. The Serpent went on to say that they would not surely die and the reason that God did not want them to eat of the tree is that they would become like Him, knowing good and evil. By then Eve's view of God was that He was overly strict and was holding out on her. Now comes the essence of the temptation. She saw that it was good for food, to please the flesh; she saw that it was pleasant to the eyes, the gate to the soul; and desirable to make one wise, an appeal to her spirit.

Notice this threefold temptation and how Satan will use this same approach on the Lord in Matthew 4. The first temptation was to turn the stones into bread, to please the flesh; then to jump off the temple, this would show that He is God as He floats down to the bottom and would please the soul; then to bow and worship the Devil to receive the kingdoms of the earth, to pollute His spirit. But our Lord would have nothing to do with the temptations of the Devil but rather responded with the Word of God. In I John 2:16 we are given a warning concerning this threefold temptation, "For all that is in the world—the lust of the flesh, the lust of the eyes, and the pride of life—is not of the Father but is of the world."

In this way sin entered into the world. Apparently Adam was not far away for as soon as she had eaten of the fruit she turned to her husband to give him a bite and he ate as well. As already noted in the above paragraph, it is Adam who is responsible for what has taken place as he is seen as the first Adam who disobeyed and brought sin and death into the world. Our Lord Jesus is seen as the last Adam who obeyed God and brought salvation and life Romans 5:12-21. Sin has now passed to all of mankind in that all now sin. It is the natural bent of humanity to sin from the womb forward Psalms 51:5; 58:3. To put this to the test just ask yourself when you learned how to sin or when was the last time you saw a parent sit down with their children to teach them how to cheat, lie, disobey their parents etc. Rather it is always the reverse that it is true, we have to teach our children how not to sin or how to be good.

This is what is known as the total depravity of man. As we read on in the book of Genesis we see how this plays out. In chapter 4 we have the first murder as Cain killed his brother Abel and then went on to build a civilization filled with murder. In chapter 6 "God saw that the wickedness of man was great" and the result was the flood, but "Noah found grace in the eyes of the Lord". It was not long after the flood that mankind gathered together to build a tower for the worship of the heavens and God confused their languages and scattered them over the face of the earth. In chapter 19 we have the account of Sodom and Gomorrah where their sin was so great that God could not find 10 righteous people to save the cities and they were destroyed. It is all summed up in Romans chapters

1 thru 3 where we have the doctrine of the total depravity of man. In Romans 1 we have the threefold result of the threefold temptation. In verse 24 "God also gave them up to uncleanness, in the lusts of their hearts, to dishonor their bodies among themselves"; in verse 26 "God gave them up to vile passions."; and in verse 28 "God gave them over to a debased mind". In chapter 3 "we have that there is none righteous, no, not one;" (Vs 10) and "there is none who understands; There is none who seeks after God." (Vs 11). The text goes on to establish that no one is righteous before God, all have failed and therefore all are in need of salvation by faith in Christ.

So how many times does a puppy have to bark to become a dog, once or twice? The answer, of course is none, a puppy barks because he is a dog not to become one. The same thing is true of us, we sin because we are sinners not to become sinners. This sin has also transferred to all of creation. When God came looking for Adam to give him the opportunity to step out in the open and confess what had happen, Adam was in hiding, afraid of his nakedness. When he finally stepped out from hiding the divine interrogation began. God asked, "Who told you that you were naked? Have you eaten from the tree of which I commanded you that you should not eat?" Adam then blamed God and his wife for his sin, it was the woman you gave me who deceived me. The woman then blamed the serpent and the serpent had no where to turn. Hence, it has been ever since that men and women constantly try to side step their personal responsibility before God to either obey or disobey.

Then came the judgment of God. God is love, grace and mercy but He is also holy and just. Sin will always have consequences with the ultimate consequence being death. The serpent was cursed to crawl on its belly in the dust of the earth and there will be a fear between the offspring of the serpent and the offspring of the woman. In this curse there was also a blessing, that even though the serpent would wound the seed of the women that seed will bruise his head. In this is the first prophecy of salvation. The woman will bear children but it will be in pain and sorrow and the man shall rule over her. To Adam He said that from now on he will till the ground by the sweat of his face and the ground will be cursed with thorns and weeds. Then after all the labor of ones life, from dust you came and to the dust you will return. In this curse of the ground we have the consequence of sin on the earth. Therefore, we have both moral sin (evil) which is disobedience to God and natural sin (evil) which is a result of the curse. So when a tornado rips through a city and destroys life and property it is a result of a sin cursed earth.

The ultimate consequence of sin is death. To understand what happen in Genesis we need to understand what the word *death* means. The word means separation so that spiritual death is separation from God; physical death is the separation of the soul and spirit from the body. When God said in Genesis 2:17 ". . . for in the day that you eat of it you shall surely die.", He was warning of spiritual death in that man would be separated from God and physical death in that from that point on mankind would be dying until the final death. In other words the translation could read, "in dying you will die." There is a point in which our human bodies begin to wear out and begin the journey toward death and in creation everything is moving in the direction of decay. Therefore we read in Romans 6:23, "For the wages of sin is death . . ."

So what we have so far is that there is one God and the one humanity is separated from the one God by sin. The way back to the one God is through the one mediator, our Lord Jesus Christ and His one sacrifice for sins and then His resurrection. To complete this first chapter we now need to review how we appropriate this one way to the one God. This will also be a word study in defining these words common to our salvation.

Our primary text is Romans 3:21-25 where we will pick up on these words as they appear in the text. The first one being **righteousness** in verse 21, "But now the righteousness of God apart from the law is revealed,". In a later chapter on apologetics we will more thoroughly discuss the topic of right and wrong and the many theories on how we arrive at what is right or wrong. For our purposes in this chapter we will simply define righteousness as what is right according to the standard of God. This definition is based upon the premise that there must be an absolute standard of what is right and therefore whatever does not meet that absolute standard is wrong. In this definition God is His own standard of what is right and as an absolute being His standard is absolute.

The next word is **faith** in verse 22, "even the righteousness of God, through faith in Jesus Christ,". The most basic meaning of the word is to commit to something or someone. In the Greek text the root word for faith is the same as for believe and trust. Based upon Hebrews 11:1, **faith can be defined as believing in that which you cannot otherwise attest to with the 5 senses**. This stands in contrast to what is considered to be modern science in that all things that exist must be proven empirically i.e. in the realm of the physical. But even scientist must agree that there are some things that cannot be proven in the physical realm and yet they do exist such as human emotion. For example, you cannot take the human emotion of joy and explain it based upon empirical evidence. You can see the evidence of joy on ones face etc. and in the same manner you can see the evidence of our faith in creation. It is more logical to assume that the things which we see were made by an ultimate uncaused cause even if that cause cannot be seen.

Next is the word **sinned** in verse 23, "for all have sinned and fall short of the glory of God,". This is a word that simply means to miss the mark of God's standard of righteousness. No matter how hard we try we cannot help but miss God's standard and to fall short of His measure. Some may do better than others but the simple truth is that we all miss and therefore all have sinned.

There are three words that follow in verse 24, **justified**, **grace** and **redemption**, "being justified freely by His grace through the redemption that is in Christ Jesus,". **Justified** is a judicial term and therefore is a judicial declaration of righteousness based upon our faith in Christ Jesus. It is not "just as if I never sinned" it is because we sinned we could not save ourselves but because of our faith in the Lord Jesus we have been declared righteous by the righteous judge. **Grace** is an attribute of God whereby He bestows a favor upon His enemies, that favor being His mercy. Grace is translated from the Greek word charis which is an attribute of something or someone that gives joy to the beholders of it. We get our English word charisma from this word and while we may say that a person has charisma we do not really know what that is, we just know that there is something about that person that gives joy to those who behold him or her. The fact that God has bestowed a favor upon His enemies is beyond what is seen in Greek philosophy as the Greeks would never think of someone doing a favor for an enemy. This is why Romans 5 is so important in explaining that Christ died for the ungodly (vs 6). The word **redemption** comes from the word redeem which is to buy back. If we should pawn a watch, for example, to buy it back is to redeem it. We who were lost have been bought back by the purchase price of the blood of Christ.

This brings us to the word **propitiation,** "whom God set forth as a propitiation by His blood,". The simple meaning of the word is to cover. In the Old Testament it was as when the blood covered the mercy seat then God was satisfied. Therefore, it is that which eternally satisfies the just claims of a holy God.

This can be all summed up in Ephesians 2:8-9 where we read that it is by grace that we have been saved, that is our salvation has originated from the attribute of God which is grace. Then it is through the channel of faith or the means to appropriate the grace of God for ourselves is by faith or the commitment that we make to that grace and the person of our Lord Jesus Christ. It is not of ourselves in any way and there are no works that we can do to merit God's favor, it is totally a gift that is given by God to us. Since it is a gift it cannot be of works because then it would be a payment of a debt or wages that are owed. Since it is not the payment of a wage it must be a free gift (as per Romans 11:6).

Since our salvation comes from God then it cannot be lost. This is why it is called **everlasting life John 3:16** rather than maybe, hope so life. We have the assurance of our Lord, **"And I give them eternal life, and they shall never perish; neither shall anyone snatch them out of My hand." (John 10:28)**. Then we have the great assurance passage of **Romans 8:28-39** where we read that it is not possible for anything to separate us from the love of God. Inasmuch as our eternal security is so important, the next chapter will be given to it, to explain further.

Read more about it:

Basic Theology by Charles C. Ryrie
Lectures in Systematic Theology by Henry C. Thiessen
The Genesis Record by Henry M. Morris
Commentary of the Whole Bible by Jamieson Fausset and Brown
Englishman's Greek Concordance Zondervan Publishing
Thayer's Greek-English Lexicon by Joseph Henry Thayer
The Analytical Greek Lexicon Zondervan Publishing

The Greek New Testament edit. Kurt Aland, Matthew Black, Carlo M. Martini, Bruce M. Metzger, and Allen Wikgren

The Interlinear Literal Translation of The Greek New Testament by George Ricker Berry

Word Studies in the Greek New Testament by Kenneth S. Wuest

CHAPTER TWO

ETERNAL SECURITY

Over the years there have been times when after people hear my teaching on evangelism, they will say, oh you believe in once saved always saved. I am certainly glad that they come to that conclusion after hearing my teaching but I cannot help but wonder why would anyone not believe once saved always saved? In the previous chapter I outlined our great salvation. Further study on this subject would include a study of the book of Romans which sets down the argument for our salvation as a legal brief might be constructed. After giving our attention to any study of our salvation, regardless of how brief it would be, how is it possible to trivialize it down to the point where it can be seen as something that can be lost? A person might lose their car keys or other such material things, but to say that you can lose your salvation is to prove your ignorance not only of soteriology but of theology proper. The intent of this chapter is to examine the biblical text having to do with our eternal security and to examine the arguments given and the texts that are used to support the arguments for losing one's salvation.

There are an abundance of texts which teach that our salvation is secure in Christ, therefore it will not be possible to examine them all. There are two books of the Bible which give a great deal of teaching on the subject and so more time and space will be spent in those two books than others, though others will be cited as well. The first is the Gospel of John. It is this Gospel that teaches more on our salvation than does any of the other Gospels. It is important to keep in mind that the evangelist in this Gospel is none other than our Lord Himself. As it pertains to our eternal security the key word is αιωνιος which is translated either *eternal* or *everlasting*. It appears seventeen times in the Gospel of John, more times than in any other New Testament book. The meaning of the Greek word is just as it is translated i.e. everlasting, eternal and perpetual. The word is used for both everlasting life in Christ and in glory or everlasting punishment. It could be argued that all mankind lives forever whether in constant bliss or constant punishment, therefore the Lord was only saying that a person will live after death, not necessarily in heaven. Therefore, it becomes necessary to distinguish between the two which will be done by examining the context.

In the Gospel of John the first mention of the word is in John 3:15, ". . . that whoever believes in Him should not perish but have **eternal** life" which is immediately followed by the most well known verse in the Bible John 3:16, "For God so loved the world that He gave His only begotten Son, that whoever believes in Him should not perish but have **everlasting** life." The conversation with Nicodemus begins with the principle of the new birth. A person must be born again to **see** the kingdom of God (vs 3) and they must be born again to **enter** the kingdom of God (vs 5). The Lord emphasizes the necessity of the new birth in verse 7 where we read, "Do not marvel that I said to you, You must be born again." There is an interesting part of the grammar of this verse that is worthy of note. When our Lord tells Nicodemus do not marvel, it is a singular noun so that it is Nicodemus who must not marvel. But when He says you must be born again it is a plural noun. It is not only Nicodemus that must be born again but everyone throughout all time in every people group around the world who must be born again. This relationship that we have with the Lord Jesus is a birth relationship. We have been born into the family of God, we are His children. In I John 3:1 we read, "Behold what manner of love the Father has bestowed on us, that we should be called children of God!" The word *children* is translated from τεκνα which is literally *born ones*. It is important to think this through. As those who have been born into a family, we cannot be anything else but children. You cannot divorce or disown your children, regardless of what they do, they are still the child of their parent. A child will take on the characteristics of their parent. This is exactly what the Scriptures teach, "But we all, with unveiled face, beholding as in a mirror the glory of the Lord, are being transformed into the same image from glory to glory, just as by the Spirit of the Lord." (II Corinthians 3:18). And ". . . but, speaking the truth in love, may grow up in all things into Him who is the head—Christ—from whom the whole body, joined and knit together by what every joint supplies, according to the effective working by which every part does its share, causes growth of the body for the edifying of itself in love." (Ephesians 4:15-16). Our relationship with God is not based upon law or upon any kind of *set of rules* or *traditions* or *sacraments* etc. where if we fail to keep it, we are no longer a child of God. Once you are a child, you are always a child.

This is important to understand because a relationship based upon the keeping of the law will always fail. As we read in Romans 8:3, "For what the law could not do in that it was weak through the flesh, God did by sending His own Son in the likeness of sinful flesh, on account of sin: He condemned sin in the flesh,". This is the argument made throughout the book of Romans, that it is not possible to keep the law or any other type of ruling principle because of the weakness of the flesh. In this flesh we simply cannot do it, we cannot measure up. This is particularly true of God's standard of righteousness but is also true of any man made standard of righteousness. All that we can do is the best that we can and that is not good enough. In contrast our birth relationship is one where we are growing and maturing in the things of God, as a child of God, day by day and moment by moment through the working of the Holy Spirit. It is like the old saying, "Do this and live, the law demands, but gives me neither feet nor hands. A better word the gospel brings. It bids me fly and gives me wings."

This brings us to the question of how do we become born again and gain this birth relationship. Our Lord is very clear in His answer that everyone who believes in Him will have everlasting life. Again, we have an interesting grammar here that is worthy of note. The word *believe* is translated from πιστευων which is a present participle. This means that it describes both the action and the person. In other words this is a believer who believes. The early Church understood this as they called one another believers. It is the Greeks who called them Christians and it was not necessarily a compliment. So then this is a person who believes and then keeps on believing and the reason why they keep on believing is because they are a child who has been born into the family. Those who have been born into the family now have a new nature which gives them the inclination to keep on believing. Then there is the verb *has* which is translated from the word ech which is a present subjunctive mood which means that the action of the verb is contingent upon certain existing and known conditions. In this verse the existing and known condition is that the person who is a believer will have everlasting life. Therefore, the everlasting life spoken of here is everlasting life with Christ in His glory. Not just anyone can have this everlasting life and not everyone will have this everlasting life, it is only for those who are believing in Him. There is only one way to the one God.

This truth that those who believe have everlasting life is emphasized in verse 36 where we read, "He who believes in the Son has **everlasting** life; and he who does not believe the Son shall not see life, but the wrath of God abides on him." Here is the contrast between those who believe in Him have everlasting life with Him, but those who do not believe will suffer under the wrath of God. There are people who ask why God does not do something about evil? One day they will discover that God has done something about evil and will again do something about evil. Those who believe will live on in eternal bliss while those who do not believe will live on in eternal damnation.

In John 14 our Lord is speaking to the woman at the well. He began the conversation with a request for water. Our Lord is the master at turning conversations from physical things to spiritual things. The woman asks why He would be asking for a drink from her and the Lord replied, ". . . but whoever drinks of the water that I shall give him will never thirst. But the water that I shall give him will become in him a fountain of water springing up into **everlasting** life." Our Lord is fond of using water as a metaphor for the life that we have in Christ through the indwelling Holy Spirit. He does it again in John 7:37-38, ". . . If anyone thirsts, let him come to Me and drink. He who believes in Me, as the Scripture has said, out of his heart will flow rivers of living water." The way in which water works for us is that we drink it, we take it in. In the same way we take in the Holy Spirit by believing on the Lord Jesus. The result is not that the water comes and goes but that it is a fountain of water and rivers of water which flow forever into everlasting life. This is the work of the Holy Spirit within us and it is a work that cannot be undone by anything that we do because we are not greater than the Holy Spirit.

As this story continues in John 4, the disciples return from town with food and ask the Lord to eat but He replies that He has food that they do not know about. He then goes on to instruct them to lift up their eyes to fields that are ready for the harvest. The harvest is to bring in fruit for eternal life, ". . . lift up your eyes and look at the fields for they are already white for harvest! And he who reaps receives wages, and gathers fruit for **eternal** life, that both he who sows and he who reaps may rejoice together." (John 4:35-36). This is another common metaphor used in the Bible for the preaching of the gospel with the result that people will believe on the Lord Jesus and will have eternal life. It is the principle of sowing and reaping where the gospel is the seed that is sown, it then takes root in the heart of those who believe it and then those are the fruit for eternal life. We see the same picture given in I Corinthians 3:6-8 "I planted, Apollos watered, but God gave the increase. So then neither he who plants is anything, nor he who waters, but God who gives the increase. Now he who plants and he who waters are one, and each one will receive his own reward according to his own labor." Here we have the teaching that there might be several people involved in the presentation of the gospel, the end result is that salvation is of the Lord and He gives the increase. The point here is that there is to be a harvest to eternal life, not maybe, hope so, if all goes well life.

In John 5:24 we read, "Most assuredly, I say to you, he who hears My word and believes in Him who sent Me has **everlasting** life, and shall not come into judgment, but has passed from death into life." Our Lord has just made statements that He is equal with God His Father. The Jews were ready to stone Him. He then makes the statement that anyone who hears His Word and believes on the one who sent Him has everlasting life. Since He and the Father are one believing on the one who sent Him is equal to believing in Him. The result is everlasting life, a person passes over from a state of death (separation from God) to a state of life (born into the family of God). Inasmuch as God is eternal, this life that He gives is also eternal.

Our Lord continues His argument in John 5:39, "You search the Scriptures, for in them you think you have **eternal** life; and these are they which testify of me." Searching the scriptures and finding eternal life is to find our Lord Jesus. Our Lord Jesus is God and being God He is eternal and therefore this life that is in Him is as eternal as He is eternal. There are no on going conditions for Him or for this life that He gives.

In John chapter six our Lord will launch into His great discourse on the bread of life. This occurs on the heels of the feeding of 5,000 men and His walking on water to cross the Sea of Galilee. Many of the people who had been fed followed in boats and asked how He had gotten across the sea because they knew there were no more boats and they had seen the disciples leave without Him. Our Lord started out with a rebuke that the only reason they followed Him because they ate of the bread that He had provided. Our Lord will now move their attention from the physical to the spiritual. They labor for physical food but their priorities are wrong, they need to labor for spiritual food. The physical food will perish but the spiritual food will endure to **everlasting life**. At this point it may appear that one must labor to have everlasting life or that our salvation is by works. Again, we must allow the context to instruct us. The point here is that this salvation is to be pursued with the same passion and diligence as one pursues his daily work for physical things. We see this in verse 27 where when one has pursued **everlasting life**, it will be *given* to them by the Son who has the Father's seal on Him. The Father and Son are together in giving **everlasting life** to be sure that it is everlasting. It is the will of the Father that whoever believes on Him may have **everlasting life** verse 40, therefore whoever believes on Him has **everlasting life** verse 47. It ought to be abundantly clear at this point that God (Father, Son and Holy Spirit) has determined that those who believe on the Son may have everlasting life based upon the fact that they have believed on the Son. This everlasting life is by the will and purpose of God and is not subject or influenced in any way by the actions of those who believe, once they have believed.

In chapter ten we have the discourse of the great Shepherd. This chapter contains two of the seven great *I am* statements of our Lord: "I am the door" (vs 9) and "I am the good shepherd" (vs 11). The good shepherd gives His life for the sheep, the sheep know Him and He knows them, the sheep follow Him because they know His voice, and He gives them **eternal life.** The reason this is eternal life is because it is given by the Lord and He holds them in His hand and no one can snatch them out. The eternal security of the believer is totally dependent upon the ability of the Shepherd to hold on to His sheep.

In John chapter twelve our Lord begins to teach concerning His death and resurrection. He refers to a grain of wheat that falls into the ground, dies and then raises up to produce much fruit. He then taught that to love this life is to lose it, but to hate this life is to keep eternal life. This life is that which can be lost because we have eternal life which cannot be lost. It is the command of the Father that we have everlasting life (vs 50). The end result of our Lord's life, death and resurrection is that we have everlasting life.

Finally, in our Lord's high priestly prayer in John seventeen He began with how all authority has been given to Him that He should give eternal life to all those that the Father has given to Him. It is important to note that once again the giving of eternal life is the work of God, not of man. He goes on to pray that this eternal life is to know God and our Lord Jesus Christ whom He has sent. To know God is to have eternal life and to have eternal life is to know God. There is no where in any of these verses that even intimate that a person can lose their salvation through their lack of effort or obedience. Under the law there were definite statements that if they obeyed the law they would be blessed but if they disobeyed they would be cursed. There are no such statements made concerning our salvation and eternal security.

It would be well within the mark to say that the chief argument against eternal security is that this gives the Lord's people permission to live any way in which they want because they have nothing to lose. This view ignores the birth relationship that we have in Christ as explained earlier in this chapter. We are His children and as we mature we take on the characteristics of our father. We live holy lives because of who we are in Christ not to attain to our salvation or to maintain our salvation. This issue is addressed specifically in Romans 6.

In Romans 6:1 we have the question, "What shall we say then? Shall we continue in sin that grace may abound?" The answer is given immediately in verse 2, "Certainly not! How shall we who died to sin live any longer in it?"

The emphasis is placed upon our birth relationship. The old man or that which we were has been put to death through our faith in Christ. Believing is more than just mental assent, it is a full commitment to and identification with the death and resurrection of our Lord. As our old man died with Christ, the new man has been resurrected with Christ. This is yet another way of describing what is meant by the new birth. This is described for us in terms of baptism (verses 3-6). The word *baptize* comes from βαπτιζω which is literally *to dip* or *submerge*. Therefore, to be literal, to sprinkle is not baptism because to baptize is to submerge. The picture being painted here is that just as we are submerged into the water we have been submerged into the death of Christ. Just as we raise up out of the water, we have been raised up to this new life in Christ. This life that we now live, we live according to this new nature having been set free from the bondage of sin. The result of this new life in Christ is that there is now no condemnation, "There is therefore now no condemnation to those who are in Christ Jesus." (Romans 8:1). This could not be accomplished by the law, because the law was weak through the flesh (8:3). Simply stated, a person cannot meet the just demands of the law in the flesh, they must be born of the Spirit to be set free from the bondage of sin so that they now serve God through the indwelling Spirit of God. This service to God and this holy life is not so that we might maintain our salvation it is because of this new nature working itself out in us.

After we believe on the Lord Jesus, the Holy Spirit enters into us and raises us up to this new life. Therefore, we should expect to see righteous living coming from the inside out. We see this theme in Philippians 1:6, ". . . being confident of this very thing, that He who has begun a good work in you will complete it until the day of Jesus Christ;". Then again in chapter 2:12,13, "therefore, my beloved, as you have always obeyed, not as in my presence only, but now much more in my absence, work out your own salvation with fear and trembling; for it is God who works in you both to will and to do for His good pleasure." At first glance it appears that we are to work out our salvation. We already know that in the flesh we are without the ability to consistently live a righteous life. This is why it is so important that we keep reading, because it is God who is working out our salvation within us. This is done by the indwelling Holy Spirit who is God. The doctrine of the trinity becomes very practical here as it is God the Holy Spirit who is working out our salvation which is the work of God the Son which is offered up to God the Father as a sweet offering of love and devotion.

Moving on in the book of Romans we have the great passage on our eternal security in Romans 8:35-39:

Who shall separate us from the love of Christ? Shall tribulation, or distress, or persecution, or famine, or nakedness, or peril, or sword? As it is written: "For Your sake we are killed all the day long; We are accounted as sheep for the slaughter." Yet in all things we are more than conquerors through Him who loved us. For I am persuaded that neither death nor life, nor angels nor principalities nor powers, nor things present nor things to come, nor height nor depth, nor any other created thing, shall be able to separate us from the love of God which is in Christ Jesus our Lord.

This text begins with the question, "Who shall separate us from the love of Christ?" the answer has already been provided in previous verses 33 and 34. There the questions were "Who shall bring a charge against God's elect?" and "Who is he who condemns?" There is no one and the only one who is qualified is the one who died for us and He will not because He is making intercession on our behalf. There is no person but is there a thing that can separate us from God's love? The answer to that is nothing as seen in the list given in verses 38 and 39. The summary is that there is no created thing that shall be able to separate us from the love of God which is in Christ Jesus our Lord. Some have argued that it is true that nothing else can take away our salvation but that we can lose it ourselves. This is contrary to this text in that there is no created thing, included ourselves who are created things, that can separate us from the love of God. There is an on going struggle here between man and God as to who is in charge.

In the first chapter of Ephesians we have dogma that teaches who is in charge of our salvation and who is responsible for our security in Christ. Ephesians 1:3-14 is divided into three parts. The first part is what God the Father has done in verses 3-6; the second part is what God the Son has done in verses 7-12; and then the third part is what God the Holy Spirit has done verse 13-14.

What God the Father has done is in verses 3-6 and include the following:

- Blessed us with every spiritual blessing in the heavenly places in Christ.
- He chose us in Christ before the foundation of the earth
- Predestined us to the adoption of sons

This is done by grace without any merit on our part either before or after believing faith. God has seated us in the heavenlies in Christ (see also 2:6). He first chose us so that we will chose Him and then set us within boundaries so that we will hear the gospel, understand the gospel and then believe the gospel. This was all done to the praise of

His glory and there is nothing in this that we can claim to be our own, it is God and God alone.

What God the Son has done is in verses 6-12 and include the following:

- Redeemed us by His blood
- Forgiven our sins
- Made known to us the mystery of His will
- Will gather us all together in the dispensation of the fullness of times
- In Him we have obtained an inheritance

We have been redeemed and forgiven by the grace of God through the predetermined act of God. The Son of God has accomplished the great salvation with the end result that we are all made one in Him and are joint heirs with Him. This has nothing to do with what we have done or will do in the future, it is all of God. This to the praise of His glory.

What God the Holy Spirit has done is in verses 13-14:

- Sealed us with the Holy Spirit
- Given us the earnest of our possession

The sealing of the Holy Spirit does two things. It shows ownership and it provides security. The seal is that of a king and it cannot be broken, therefore, those contained within the seal are forever secure. What God does cannot be undone by man. The Holy Spirit is our earnest which is guarantee that what we know of the Holy Spirit today is as good as or better than what we will know in glory. In other words, if you think that what we have in Christ through the working of the Holy Spirit is good today, then you have not seen anything yet. The best or the fullness of what we have is still to come. Our possession in glory is secured by God and it is to the praise of His glory.

It ought to be clear that this great salvation is the work of God, Father, Son and Holy Spirit and is to the praise of His glory. Therefore for any man or woman to say that they can some how lose it is the height of human presumption and arrogance. Let us rather bow done to His sovereignty and be thankful.

There are those who struggle with Hebrews 6:6 where we read of the *falling away*. It is not possible to give full treatment of this text here but to only point out certain salient points. First, this text is not dealing with eternal security or lack thereof, but rather is emphasizing the all sufficiency of the sacrifice of Christ. The text is framing an hypothetical situation to show that if it were possible to "fall away" then there is no other sacrifice for sins. If the one sacrifice of our Lord is not sufficient to forever take away sins then there is no other (see Hebrews 9:26-28; 10:10-12). This then leads us to the second point which is if it is possible for one to "fall away" that person can never come back, 'For it is impossible . . . if they fall away, to renew them again to repentance." (verses 4-6). If you can lose it, you will never get it back. The point being is that you cannot lose it because our Lord's one sacrifice for sin is sufficient for all time. The sacrifice of animals could not accomplish this, but the superior sacrifice of our Lord permanently and perfectly satisfied the just claims of a holy God. There is no more sacrifice for sins because no other sacrifice is needed. Therefore, once we are saved, we are forever saved.

In summary it is important to remember that our salvation is based upon the work and grace of God not upon any human merit either before or after the point of our salvation. We do not have the power to overthrow the work of God. When we get to heaven we will have no other merit than the grace of God and will have only Him to honor and to worship.

CHAPTER THREE

PSYCHOLOLGY OF EVANGELISM

When I was a young man an older brother in the Lord was teaching me how to do door to door evangelism. He told me to picture in my head that I had two six guns in my holster and when the person opened the door, I was to draw one of my guns and start shooting. But I was to be careful not to shoot all my bullets at once, but to save some in the event I needed them later. This is a mental picture that I have never forgotten as it speaks to the psychology of evangelism, i.e. how do we communicate the gospel to all the different kinds of people that we meet. For some, just a few shots will get us a hearing, for others we will need to empty both barrels. So it is important to think about our delivery of the gospel.

Over the course of years I have studied styles of evangelism. Among the styles of evangelism there are basically two i.e. Life Style and Confrontational. In his book *Life Style Evangelism* Aldrich states, "The thesis of this book is that a Christian becomes good news as

Christ ministers through his serving heart. As his friends hear the music of the gospel (presence) they become predisposed to respond to its words (Proclamation) and then hopefully are persuaded to act (persuasion)."[5] This is the most popular style of evangelism because it is the natural outgrowth of our godly behavior and life. Floyd McClung said, "People don't care how much we know until they know how much we care."[6] This simply means that as we love and care for people then they will be drawn to us and to our life as a believer. It is important to keep in mind that in order for this style of evangelism to actually reach people there has to be a point somewhere in the relationship where the Gospel is actually presented.

Another style of evangelism is not so popular particularly in this day and age. I mean by that there was a time when there were evangelists who were very aggressive, even to the point of running people down and tackling them. Those folks are asleep in the Lord and have since gone on to their reward. To confront has been defined as to stand facing or opposing in a challenge, defiance, or accusation. This style was actually very effective as the old time evangelists led people to Christ on a regular basis. Obviously the popularity of this style has waned

---

5   Joseph C. Aldrich, *Life-Style Evangelism* (Portland: Multnomah Press, 1981) 81.

6   Ibid. 35

somewhat over the years as most modern day Christians are concerned about offending people.

I have preferred a style that I call *Attack Evangelism*. I think I made it up but since I was born too late to be original I do not know that for sure. I define it as "an evangelistic mental attitude that is on the offense without being offensive. A mind-set that will recognize and take advantage of opportunities and will create opportunities where none had existed." Note the words a "mind-set" or an attitude. This is intended to bring together the best of both the other styles in that as we live out our lives in a godly manner, we look for opportunities to give the gospel. Then as these opportunities present themselves, we take advantage of the opportunity to give the gospel. Then we also look for ways to create these opportunities through ministry. While serving others we again look for opportunities to give the gospel. In looking for opportunities, we are being aggressive rather than passive. We are going to be on the offense without being offensive.

Even with *Attack Evangelism* building relationships is important. Therefore we need to discuss what exactly this means? Larry Moyer gives 4 B's[7] to keep in mind.

First, be accepting. The Lord does not always put people in our path who are just like us or are part of our

_____
7   Larry Moyer, *Non-threatening Evangelism Training System* (Dallas: Evantell Inc, 1992) 27.

same social, economic, status. In fact, many times the Lord will bring us someone we do not even like (who says God does not have a sense of humor). We must always remember that the cast off of society were the most attracted to our Lord while He was on the earth, it was the religious people who criticized Him for eating and drinking with sinners. His response was that the sick were the ones in need of a physician. Oddly the casts off of society today are the least likely to come to Church or to feel welcome should they attempt it. It is good to keep in mind that for the grace of God there go I. We must accept the sinner where we find him/her and remember the greater the sinner the more likely they are to listen to our good news.

Second, be a good listener. In conversations we tend to want the other person to stop talking so that they can marvel at our intelligent speech. In sales we are always taught to listen to people and they will show us how they can be sold. The same is true in evangelism, the best way to understand how to proceed is to listen as people will invariably tell us where they are coming from and what is most troubling to them. As we ask questions be sure and listen to their answer intently so that we can show sympathy and provide direction toward the good news of how the Bible has the solution to their chief concern.

Third, be complimentary. This is particularly true when a person brings up an objection. A person may say, "I don't believe the Bible is true" and our first response is a compliment such as "obviously you have given this a lot of thought" or "I can appreciate that, others have had the same problem at first". Then we follow with a question, "What is there about your life that brought you to that conclusion?" Never use the words "yes but" as this starts an argument. Use a compliment as a cushion to continue with the good news.

Fourth, be transparent. Do not try to be something you are not particularly when working with people from another culture. If you are white do not try to sound black, they will smile nicely but will be laughing on the inside. Be yourself and if you do not know the answer to a question, do not be afraid to say so and that you will get back to them with the answer. If you feel nervous, admit it. People prefer honesty over a slick sales pitch, so just be who you are, a sincere person, just sharing good news.

Before we can build a relationship we have to find the people with which to build the relationship. There are many ways to contact people and give the "good news". The most direct approach is to go through the neighborhood and knock on doors and say, "Hi, we are a group of Christians in the neighborhood and we are here to show you from the Bible how you can know for sure that you are going to heaven, may we do that?" A less direct approach is to use a survey from the Church where we want to know how we can be of service to this community. This tends to be a bit intimidating for everyone concerned so it is usually easier to develop relationships with non-believers.

Some ideas for developing relationships can include such things as:

- Join the local YMCA or other recreational center
- Participate in neighborhood projects
- Use services provided by non-Christians e.g. plumbers etc. When you are paying someone to do a job in your home, they have to listen to you.
- Invite others over coffee or lunch. Have a dessert time for neighbors, this is less expensive and time consuming.
- Be an active member in any neighborhood associations and play an active part in the community as a whole.
- Create a ministry/service that would meet a need in the community.
- Create a relationship with neighbors and co-workers.

In other words, just as you would net-work to develop a business, net-work to develop relationships which will give you a hearing for the Gospel. Or as you net-work to develop a business, use the same network to discover divine encounters and give the gospel.

Once a relationship is in place remember the 4 B's of cultivating a relationship. Then keep in mind these basic principles. Start sharing truth early. One of the problems of friendship evangelism is that we tend to make friends but never get around to sharing the gospel. Start inserting truth statements into your conversation early in the relationship so that it will come as no surprise when you come out with the greatest truth of all. In this day and age where there are no absolute truths, truth statements will get their attention. Be patient and prayerful. More people are offended and driven away because Christians got into too big a hurry and did not allow for the Holy Spirit to work with the truths already given. Always be prepared to share the Gospel. When the opportunity presents itself present the gospel, this is not the time to shy away and later wonder why you did not go ahead and give them the good news.

Once a friendship/relationship has been established it is now time to move into the Gospel. Keep the picture of a funnel in your mind as we want to move from general things to specific things. Up till now our conversations have been more in the general but now it is time to move to the specific. We also want to move from the secular to the spiritual. We may have already dropped some hints about spiritual things by inserting truth statements in our previous conversations, now we want to deliberately move the conversation to spiritual things. This may happen during a regular conversation, at other times we can actually plan this conversation in advance. For example invite the friend over for coffee and desert. Once this has been done there are some tips to help with the conversation.

<u>Create a casual atmosphere</u>. In the same way that previous conversations were done in a casual environment it is important to maintain that atmosphere now. There is no reason to take on religious airs, continue being the friend that you have been. <u>Convey genuine concern</u>. People are not targets or a score that we are keeping as to how many we have to led to Christ, but are people with genuine concerns with which we need to show genuine empathy. Even when I go door to door, I try to maintain a casual demeanor to the point of getting invited into the home so we can sit in the living room. I usually avoid going to a dining room table because that is where people get sold stuff. But even standing on a porch it is possible to be relaxed and casual. There have been times when people will begin telling me their problems and I will take time right there and then to pray for them. <u>Assume their</u>

openness to you as a person is an open door to talk about spiritual things. Now is the time to plow and pursue8.

One way to move toward spiritual things while maintaining a casual atmosphere and showing genuine concern is to make a statement like this, "You know in our past conversations you have mentioned this problem, and after giving this some thought I was thinking about how you might be interested in the solution that I have found for that problem". At this point you can give your own testimony, never take longer than a couple of minutes as people are not usually interested in long discussions about you, they would rather talk about themselves and their problem. Should they remain interested in the conversation then move into the Gospel, "Has anyone ever taken a Bible and shown you how you can know for sure that you are going to Heaven? May I?" When we are good listeners, people will tell us what the problem is that needs solving. **The solution to all problems begins with the Gospel**.

The gospel is the same for everyone at every age group, in every culture and regardless of a person's psychological make-up. However, the manner in which the gospel is presented can and should be adjusted according to age, culture and psyche. More specifically, we should say, the context in which the gospel is presented. We can use the

8   Ibid. 39-40.

same presentation i.e. "The Good News—Bad News" but how we get to the point of giving that presentation may vary. For example, when working with children we may have a vacation Bible School where there are games, crafts, music, skits and all kinds of things going on but the presentation of the gospel is the same because the gospel is the same. There are cultures where it is offensive to come right to the point. It is considered good manners to chat awhile and then, in a round about fashion, come to the point. Some people are more analytical and need to know that the gospel makes sense, others are more emotional and need to feel something. The question then is how do we know what we are dealing with at the time? With age and culture it can be obvious but with others it can be more difficult.

The first thing that we need to understand is how we humans are made. It is obvious that there is both the physical and the meta-physical. Those who argue that there is only a physical world run into a problem with human emotion as there are such things as joy and sadness that cannot be proven physically but clearly do exist. In theology there are two positions, the first is that we are made up of two parts, the physical and the spiritual which is known as bipartite. The second argument is that we are made up of three parts, trichotomy. I prefer the trichotomy version that man is spirit, soul and body I Thess. 5:23. The primary reason for this is that to follow the bipartite or dichotomists approach it becomes necessary to divide the soul into two parts of a higher and lower. This is necessary to explain how a person can be carnal or spiritual or that a person is alive yet spiritually dead etc.

The threefold nature of man is that we have a spirit (πνευμα-pneuma) which is our level of God consciousness and of spiritual things. Then we have a soul (ψυχη— psyche) and our body (σωμα—soma). In evangelism, the first thing we need to understand is that unbelievers are dead in trespasses and sins Eph. 2:1; this is spiritual death where even though the body and the soul are functioning, there is no spiritual understanding or even desire for God Romans 3:10.11. Therefore, do not be surprised when unbelievers reject your offer of good news because they are only doing what comes naturally to them. Unbelievers will often be trapped in a life style where the feeding of the body is all that matters to them. Their soul may only be interested in feeding the body more food than it needs or alcohol or drugs etc. Or their soul may only be interested in feeding their mind with what is known as the wisdom of this world without giving consideration to the wisdom that comes from God through the Scriptures. So, how do we give good news to someone who is spiritually dead? This is where we must depend on the work of the Holy Spirit to make them alive so that they will listen to the good news Ephesians 2:4,5. This is what we call divine encounters, where the Holy Spirit has opened their hearts to hear the gospel e.g. Acts 16:14.

So that brings us to the question of how do we present the same gospel using the same presentation to different personalities? It is important to keep in mind that the same gospel works for every personality. Therefore, we start out by giving the same presentation of the gospel. Our presentation, the Good News—Bad News approach, has questions built in to help us discover where people are coming from. As a reminder, **our goal is to give people the opportunity to hear the gospel, understand the gospel and believe the gospel**. We begin with the question, "Has anyone ever taken a Bible and shown you how you can know for sure that you are going to heaven? May I?" This gives the opportunity to hear the gospel. Then our presentation provides the opportunity to understand the gospel and then our closing question, "Is there anything keeping you from trusting Christ right now?" provides the opportunity to believe the gospel. It is at this point where we will discover the type of personality we are dealing with and what is going on in a person's life that may hinder them from believing the gospel right now.

Darius Salter in his book *American Evangelism* cites three principles in the psychology of evangelism.[9] The first of these three is to have an **"Unconditional Positive Regard"** for people. This begins with our first approach to people, right on through to the end of the relationship. It is incredible how our attitude toward people comes through us and right out in the open in the look on our face, the inflection of our voice and our body language. I am always praying that the Lord would enable me to see people through the eyes of Calvary i.e. the way in which the Lord sees people. We have been conditioned by our culture to be unaware of the people that surround us every day, we live alone in a city of thousands. We need to learn to see and be aware of people and be aware of them as having value. As Salter puts it, "The ability to perceive people as a meaningful part of one's existence . . . . They are significant because God meant for them to discover meaning and arrive at a purposeful destiny." All people have value because all have been made in the image of God and all are those for whom Christ died.

As we see people as having value and treat them with love and respect, this does not excuse sin. The grace of God does not compromise the truth of God and the truth of God does not compromise the grace of God but both

9   Darius Salter, *American Evangelism* (Grand Rapids: Baker Books 1996) 151-158.

find their perfect balance in the Word that became flesh John 1:14. It is in the Son of God that "Mercy and truth have met together; Righteousness and peace have kissed" Psalm 85:10. When the woman at the well encountered the Lord she found the freedom to take responsibility for her sin and at the same time found the solution to her problem, John 4:17-29. This is what evangelism is all about, the introduction of our Lord as the one who reveals the problem and provides the solution.

The second principle is "**Empathetic Listening**". To have empathy is to be so sympathetic as to vicariously enter into the feelings, thoughts and experience of another person. To listen with empathy then, is to feel the same emotions as the other person. Too often when we find ourselves talking to someone who is actually interested in what we have to say, we are in a hurry to say it and really are not interested in what they have to say. Subconsciously we are thinking, I wish this person would shut up so I can explain the truth of the matter. Listening with empathy is important in any relationship whether it be in a marriage, family life, business or in evangelism.

There is one problem with empathetic listening which is that we risk closeness to others. For some people this is a serious problem as they cannot get close or are afraid of getting close. When you get close to someone you start feeling obligated to do something about their problems and hurts. They are hurting, so you with empathy, start hurting and now you need to do something to stop the hurting. This now draws you into paying a cost of time, energy and money. One of the first thoughts that jump into my mind at this point is how did I get myself into this mess? The answer is very simply, I dared to demonstrate the love of Christ to someone and it all began with listening.

In the field of sales, there is a strategy that seldom fails. The strategy is to ask questions and then listen carefully because a person will invariably tell you how to sell them. They will tell you what they are looking for, what problem needs to be solved, and what the product has to do to solve the problem. Evangelism is the giving of good news and one of the reasons why this news is good is because it solves problems. So as an evangelist we need to listen to find out what the problem is and then demonstrate how the gospel is the solution to that problem. This is not as difficult as it may seem because the solution to every problem is going to be the same. Everyone needs to be reconciled to God and become a new creation in Christ II Corinthians 5:17-19. The gospel is the first solution and contains the power to solve the problem.

Therefore, we first listen so that we understand and feel the problem as articulated by the person and then explain how that specific problem is solved by becoming a new creation in Christ. If we do not listen well, we can make a mistake and address the wrong problem. These old things that have dragged this person down can be done away with when they become this new person. It all begins with being reconciled to God which makes it possible for men and women to be reconciled to others.

The third principle taken from Salter's book on American Evangelism is **Loving Care.** This is based on the thinking that "People will not respond to the gospel in a vacuum" therefore such questions as, "Does anyone care for my soul? Does anyone care about my finances, my divorce, my children, and my addiction?" all become relevant to the gospel discussion. Aldrich in his book on life style evangelism states, "People are more inclined to respond to the Gospel when they understand how trusting Christ will satisfy their needs" but is this true?

This brings us to Maslow's Needs chart. This is used by psychologists and sociologists alike. This theory is based on the premise that humans will naturally seek to have basic needs satisfied first and then will proceed up to another level of need until that need is satisfied etc.

The chart looks something like this:

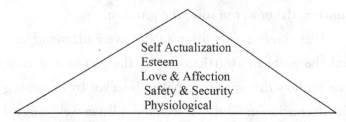

Self Actualization
Esteem
Love & Affection
Safety & Security
Physiological

The chart works from the bottom up, i.e. physical needs must be met first and then people must feel safe and on up the chart. The gospel may be presented once a number of levels of these needs are met. The question is how many levels must be achieved and at what point do we present the gospel? To make a long story short, the gospel may be presented at any time. It is important to keep in mind that the gospel is a spiritual remedy to a spiritual problem. Since the spirit, soul, and body are interconnected, a spiritual remedy will impact both emotional and physical needs. It is true that a spiritual remedy will not make a person less hungry but it will push someone who has food to share it with someone who does not. Herein lies the difference between Christianity and all other religions and philosophies. For example, you will never find a leprosy colony and hospital founded by atheists.

Our Lord certainly met physical needs as well as spiritual. When the disciples of John the Baptist asked if He was the one, his response was, "The blind see and the lame walk; the lepers are cleansed and the deaf hear; the dead are raised up and the poor have the gospel preached to them." (Matthew 11:5). We also read of Him, feeding the multitudes, healing all manner of diseases, and casting out demons. He had a great advantage over us as He is God and can do these things instantly. We can argue that we have His example to follow to do what we can do in the physical realm but must always keep before us that our main goal and first priority is to give the gospel. In fact our Lord invariably followed His miracles with the gospel.

So then meeting physical needs can give opportunity to give the gospel. We must be sure to give the gospel regardless of what level a person is on the needs chart because we do not know at what level they will trust in the Lord. Meeting physical needs will move us into the area of the Sociology of the Gospel which is our next chapter.

Taking these three principles together i.e. unconditional positive regard, emphatic listening, and loving care, we can see how it is possible to build relationships with people with a view toward giving the gospel. We can also see how this will create interesting, if not uncomfortable, relationships. Our Lord certainly exemplified all three of these principles and the result was that He was a friend of sinners. Philip Yancey in his book *The Jesus I Never Knew* pointed out, "Jesus was a friend of sinners. They like being around him and longed for his company. Meanwhile, legalists found him shocking, even revolting. What was Jesus secret that we have lost?".[10] This friendship with sinners created a great deal of friction between our Lord and the religious of His day. When they

---

[10]   Philip Yancey, *The Jesus I Never Knew* (Grand Rapids: Zondervan publishing, 1995) 149.

spoke of Him as being a "friend of sinners" it was not considered a compliment but rather a complaint.

We can take a quick review of some of the folks our Lord dealt with to get a clear picture of what we are talking about. In Luke 5:27-32 we have our Lord calling to a tax collector to follow Him. The Pharisees asked Him why He would eat and drink with tax collectors and sinners. Our Lord's answer was, "Those who are well have no need of a physician, but those who are sick. I have not come to call the righteous, but sinners, to repentance." (Verses 31-32). Somewhere in the psyche a person must acknowledge that they are a sinner or a sick person in need of a physician. During the time of our Lord the tax collector was a despised man because he was paid by taking a percentage of what was collected. There was an x amount of tax that was to be sent to Rome and whatever he collected above that amount was his. So it was possible to get rich by gouging the people. It was this kind of man that the Lord called to Himself, one who knew he needed a physician. This was Matthew who became His disciple and the writer of the Gospel of Matthew.

In Luke 7:36-50 we have the story of a woman who anointed the Lord's feet and washed them with the hairs of her head. It is pointed out in the narrative that this woman is a sinner and if the Lord were really a prophet He would know that this woman was a sinner. It can be

71

assumed that the Pharisee who made this observation, also thought that the Lord would not have anything to do with the woman. Our Lord then taught how that those who have been forgiven much will love much. Then we have the story of Zacchaeus in Luke 19:1-10, another tax collector and of a woman taken in adultery in John 8:1-11. In every situation our Lord ministered to these people, He met their needs and established how He is the solution to their problem. The key question here is how do we follow His example?

The problem with the modern day Church is that people from the low end of the social-economic structure do not feel comfortable in the typical local Church. It may not be intentional but the local Church tends to treat people differently and to attract people who are just like them. We need to be challenged with the fact that in order for a physician to treat sick people he must come into contact with sick people. Just as our Lord ate and drank with sinners we too must find ways to make contact, demonstrate the love of God and give them the answer to their problem, the gospel of our Lord Jesus Christ.

Read more about it

American Evangelism by Darius Salter
Life Style Evangelism by J. Aldrich
N.E.T.S. Non-threatening Evangelism Training System by Larry Moyer
See also training available at http://www.act111.org
The Jesus I Never Knew by Philip Yancey

Chapter Four

The Presentation

One day while going door to door in the housing projects of North Omaha I came to an apartment where I knocked and a man answered and immediately shouted at me WHAT ARE YOU DOING HERE!!! I instinctively ducked and as a matter of habit blurted out, I am a Christian in the neighborhood here to show you how you can know for sure that you are going to heaven, may I do that? The man stepped back and then said, "come on in". I went through my regular presentation and came to the closing question, "Is there anything keeping you from trusting Christ right, now?" He then replied with a question which was why was I in this neighborhood at this time? He went on to explain that He had just been praying that God would send him someone to tell him the truth and then I knocked on the door. This is what is known as a divine encounter. But even after many years of teaching the Bible and doing evangelism, I was glad that I had a set presentation to use from the time I was at the door to the closing question because he did not give me enough time to think up something to say. For us to be ready to give an answer, means just that, we need to be ready and that means having the words to say immediately on our tongue. This is the advantage of having a set presentation.

We have given some thought to our biblical basis for evangelism (Soteriology) and to how we approach people and their psychic. Now is the time to learn a presentation of the Gospel so that you can begin right away to witness to friends and family and anyone who you come across. So this chapter will be about learning a presentation that is easy to learn and to use.

If you were to ask most any Christian, "Do you know how to give the gospel?" they would say, "yes, of course". Then if you were to say, "Ok, tell me the good news, using Scripture to support your argument" it is amazing what comes out. The reason why we need a learned presentation is because even though we know what the gospel is, explaining it in terms an unbeliever can understand is another matter entirely. A learned presentation will give us confidence in giving the gospel because we already know exactly what to say and therefore, we are more apt to create opportunities for giving the gospel. A learned presentation will enable us to stay on track when people try to get us off on some other subject i.e. a bunny trail that leads nowhere. A learned presentation will keep us from saying something that is not biblically accurate. A learned presentation will help us to keep the message clear and easily understood. A learned presentation will keep us from babbling on until the person gets bored and ends the whole conversation.

Therefore, I like to keep in mind the 4 B's of a presentation, taken from Larry Moyer's Non-threatening Evangelism Training System by EvanTell.

➢ **Be Accurate**—Use terms the Bible uses in their right context. Many well meaning Christians use terminology that is not in the Bible or is taken out of context. The result is confusion on the part of the unbeliever, what is it that God is asking the unbeliever to do?

➢ **Be Clear**—Stay away from using terminology that only Church people understand. Use illustrations and examples that are common to the experience of most of the people.

➢ **Be Simple**—The gospel is a simple message, resist the temptation to start at the beginning of the Bible and work your way through or feel the need to impress someone with your knowledge of systematic theology.

➢ **Be Quick**—Even in this day and age of time/labor saving devices, people all feel rushed, like they don't have any time. You should be able to give enough information in 5 minutes for a person to make a decision to trust Christ. They may have questions that will make the time longer but that is up to them.

Before getting to the presentation let us give some thought to the importance of asking questions and we use questions to move into the presentation. By asking questions we can be on the offense without being offensive. The person asking the questions controls the direction of the conversation, the person answering the questions is always on the defensive and must wait for direction from the next question. Therefore, to engage in a conversation about spiritual things we must ask questions. As Larry Moyer teaches, we want to move from general physical things to more specific spiritual things. We may begin with such things as "How are you? Where do you work? How many in your family? Etc." to "Have you been giving thought to spiritual things lately?" When a person asks us a question with just a whisper of spiritual things we can ask, "Do you ask that because you are interested in spiritual things?" Another widely used question is "Should you die tonight and God asks you why should I let you into my heaven, what would you say?" These questions always lead us to the lead question for our presentation,

"Has anyone ever taken a Bible and shown you how you can know for sure you are going to heaven? May I?

Now for the presentation that you can easily learn, deliver and teach to others. Keep in mind that this is the presentation taught in Larry Moyer's Non-threatening Evangelism Training System by EvanTell with slight modification by yours truly.

The presentation is the good news bad news approach. We begin with a question.

Has anyone ever taken a Bible and shown how you can know for sure that you are going to heaven? May I? Answer is yes you may.

Notice that we begin with the Bible. We are not giving them what our church, group or denomination believes but only what the Bible says.

Go right to the second question of "May I?" because we do not want to start our conversation with a negative i.e. no. We also want to go right to "May I?" because people have been trained to say yes to "May I?" Remember when you were a child and you asked your mother for candy, you said, "Can I have a piece of candy?" and your mother corrected you by saying, "May I have a piece of candy?" Once using "May I?" instead of "Can I?" you received your piece of candy.

The Bible has both good news and bad news, the bad news is about us, the good news is about God, we will begin with the bad news first.

Point 1 All have sinned.

Verse Romans 3:23 "For all have sinned and fall short of the glory of God"

We all think of sin as the bad things that we do, but the word actually means to "miss the mark". The mark is the holiness and righteousness of God. For example let's say we are going to throw a rock and hit the capital dome in Lincoln. You are stronger than I am so you will get closer than me but it is not a matter of how close you are, we will both miss. In the same way it is not a matter of how good we are before God, we have all missed the mark of His righteousness.

The problem with the bad news is that it gets worse!

Point 2 The Wages of sin is death

Verse Romans 6:23 "For the wages of sin is death . . . ."

We all know what a wage is, what we have earned and what we deserve. For example let's say that I hire you to paint my house and after you are done you come to collect your wages but I refuse to pay you. You would be mad because it is what you have earned. God is telling us that what we have earned, what we deserve is death and death is eternal and painful separation from God.

There is a bridge between the bad news and the good news:

God saw that we could not come to Him so He came to us.

Point 3: Christ died for our sins.

Verse Romans 5:8 "But God demonstrates His own love toward us, in that while we were still sinners, Christ died for us."

It is important to note that God's demonstration of love is seen in the sacrifice of His own Son. This is how we answer the question of how we know that God loves us. The key to this verse is seen in the little preposition "for". It means on behalf of or in the place of. For example, let's say that you are lying in a hospital bed dying of cancer. I come and take the cancer cells out of your body and place them into mind. What would happen to me? What would happen to you? Yes, I would die and you would live because I took what was killing you (cancer) and placed in my body so that I would die in your place. In the same manner, the Lord Jesus took what was killing us (sin) and placed upon Himself so that He died in our place. He paid the wages of our sin with His death.

Just as the bad news got worse, the good news gets even better!

Point 4: You can be saved through faith.

Verse Ephesians 2:8-9 "For by grace you have been saved through faith, and that not of yourselves; it is the gift of God, not of works, lest anyone should boast."

The key word here is the word "Faith" which is to believe, trust or to commit ourselves to. For example let's say that we are going to take an airplane from Omaha to Denver. We can believe all about airplanes that they can fly, that they are mechanically sound, that the pilot knows how to fly the airplane etc. but what do we have to do to fly from Omaha to Denver? Right, we have to get on the plane. It is not enough to believe about airplanes, we have to commit ourselves to the airplane so that what happens to the plane happens to us. When we commit ourselves to the Lord Jesus what happen to Him happens to us and He arose from the dead and said that "Because I live, you shall live also." (John 14:19)

So now then, is there anything keeping you from trusting Christ, right now?

It is extremely important that we learn this closing question and that we use it. So many get to this point where there is a possibility of rejection and they start roaming all over the place, asking questions like, "Do you have any questions?" "Does this all make sense?"

"You sure you don't want to know if Dinosaurs and men lived at the same time?" Do not do that! Go right to the closing question.

Is there anything keeping you from trusting Christ right now?

In asking this question we are giving the person the opportunity to trust Christ and we are giving them the opportunity to state an objection. Notice how this question invites a no answer which means yes. At this point people tend to automatically say no e.g. "May I give you a million dollars? No." But now a person has to stop and think, no means yes I do want to trust Christ. This is what we want and we want to challenge them to give a reason why they would not want to trust Christ. Should they give a reason then we will move into apologetics which will be in following chapters.

Many times a person will say no, I cannot think of a reason why I would not trust Christ. Then we tell them that prayer is talking to God, it is not the prayer that saves them but trusting Christ, so we are going to simply tell God that we are now trusting Christ. Many times we have to help them and often I end up praying for them that they are now trusting Christ. This then leads them into the wrap up.

The wrap up verse is John 5:24 where we follow-up with 4 questions:

- Did you hear the word?
- Did you believe the word?
- What do you have?
- How do you know?

The answer is that they know because the Bible says so, which brings us back to the beginning where we said that we are going to show them from the Bible how they can know for sure that they are going to heaven.

Just as we moved into the gospel presentation by asking questions, we close our presentation with a question, "Is there anything keeping you from trusting Christ right now?" At this point a person may voice an objection. There are steps that can be taken at this point to deal with the objection. It is important to keep in mind that if the person is plain obnoxious and clearly has no interest in the gospel then it is time to move on to the next person. Remember that there are close to seven billion people in the world so we will not run out of people to talk to any time soon. An excellent question to ask at this point is:

What is there about your life that brought you to that conclusion?

If the person has a serious objection and seems to have some interest then we can give a logical defense of the faith.

Whenever you get an objection i.e. a reason for not trusting Christ, remember this three-step procedure:

1. Throw down a cushion i.e. compliment the question. For example you can say "I'm glad you brought that up" or "Obviously you given this a lot of thought" or "Others have said the same thing" or " That is an interesting point of view".

2. Restate the question for clarity. You can say something like, "What I'm hearing you say is" or "So to be clear, you are saying".

3. Then ask a question because the person asking the questions controls the interview.

Many people will have one objection but they will not have two. This means that they will throw something out there that they really have not thought about. The key is to make them think about it. Since most people have never really thought about it, they will not have an answer. At this point we can answer the objection/question and go back to our closing question with "Is there any other reason keeping you from trusting Christ right now?"

There are a few people out there who will insist on their objection. At this point we will continue to ask questions to deal with their objection and to move into a system of apologetics. The questions are as follows:

1. **Is this what you are saying?** Then repeat their answer, in other words clarify their statement so that you know for sure what they are saying. It is possible to use your voice in such a way as to make their statement sound kind of silly, but you may not want to try that.

2. **Where did you get your information?** At this point we want to challenge their point of reference. We get our information from the inspired Word of God, so how does their point of reference compare?

3. **How do you know that it is true?** People challenge our position all the time, we need to challenge the position of the non-believer and make them think through what they are claiming.

4. **What if you're wrong?** I have great fun with this one, what if you're right and I'm wrong? Well nothing really happens when we die, we just die. But if I'm right and you're wrong, you will end up in hell and I will be in heaven, you sure you want to take that chance?

A typical conversation may go something like this:

Q: I don't believe there is a God.

A: What is there about your life that brought you to that conclusion?

Q: God is just a crutch for ignorant people.

A: So what I hear you saying is that God is simply the product of our imagination to answer the hard questions of life and the universe?

Q: Yes, that's it exactly.

A: Where did you get your information that supports this theory?

Q: This is what is being taught in higher education and when you are as educated as I am then you will understand.

A: How do you know the information you are getting is true?

Q: Well they all must know what they are talking about, after all look at the facts.

A: What if you're wrong? What if there is a God and one day you will have to give an account of your thoughts and actions to that God?

Again, go back to the closing question, "Is there anything keeping you from trusting Christ right now?" In terms of over coming objections there is a need to have a basic understanding of apologetics. In a later chapter I will address apologetics and philosophical theology.

However, **do not let your perceived lack of knowledge prevent you from sharing the good news with others**. Remember that most people you talk to do not have any prior knowledge of the Bible or anything related to the Bible. Therefore, you already know way more than they do, so press ahead and give them the gospel!!

Once a person has made a profession of faith, follow-up becomes important. There are times when follow-up is not possible as is the case many times with door to door evangelism or other on-the-spot times when you give the gospel the person makes a profession and then you go your separate ways. But if this is a result of a relationship which has been developed over the course of time then follow-up is possible. The first thing to do is to get the person into a Bible study. I recommend the book "Beginning Bible Lessons" by yours truly to get them started. This book is designed to go back over the gospel then to move them toward personal devotions and then to Church. Our Church will then pick them up and bring them into our community where they can experience worship, discipleship, fellowship, stewardship and outreach and thereby grow in the faith. It is important to bring them into the Church and to be confident that here they will be well cared for. Always remember that a person you lead to the Lord will then become as involved in the Church as you are. Therefore, if you believe that they should be involved then be sure to set the proper example.

Note: The presentation and various notes have been taken from Larry Moyer's lessons in the *Non-threatening evangelism training System* by EvanTell.

CHAPTER FIVE

SOCIOLOGY AND EVANGELISM

Years ago while in prayer, I asked the Lord where I should start going door to door with the gospel and upon what part of town I should focus a ministry of evangelism. The answer came immediately, if it could be shown in the inner city that the gospel changes lives and lives change society then that would get people's attention. If it could work there, it would work anywhere. So it was off to the housing projects and to the classroom to earn a Masters degree in Urban Studies. It did not take long of either one to learn that there was going to be more to it than just going door to door with the message, because once you give the gospel to a person, you are invariably drawn into their life and into their world.

Ray Bakke wrote a book *A Theology as Big as the City*. In it he wrote a chapter entitled "God's Hands are in the Mud"[11]. The main points he established in this chapter are as follows:

---

[11] Ray Bakke, *A Theology As Big As The City* (Downers Grove: Intervarsity Press 1997) 36.

- God (Father, Son and Spirit) is not defiled by contact with the physical, even contaminated or toxic earth.
- God was planning and executing both the creation and the salvation work simultaneously.
- Creation and redemption are multigenerational, and we can be realistic about progress.
- It's not a sign of a superior spirituality to work alone personally or organizationally.
- There's no reason to assume God is present on a beautiful mountaintop but absent from the city. He's there in the stuff of local ministry.

The argument made by Bakke throughout the book has to do with the "Social Gospel" that in fact the Gospel has social implications and "social work" has Gospel implications. We cannot remove ourselves from society and hope to reach that society with the Gospel.

Having used the terms "Social Gospel" it behooves me now to explain what I am talking about. At the turn of the 20th century there were organizations that sprang up with the goal of correcting social ills, such as poverty, disease, homelessness, alcoholism etc. In the United States the first Sociologists were Christians who not only studied society in this country but were determined to do something about it. Two of the best examples of what we are talking about are the Salvation Army and the Young Men's Christian Association. The Salvation Army had its beginning in England and started in the US in 1880. The 3 S's of the Army are Soup, Soap and Salvation. There can be no denying the humanitarian aid that has come from the Salvation Army, their work has been beneficial to the poor and down trodden of most every country in the world. But go up to someone ringing a bell at Christmas and ask them how they know if they are going to heaven? I would venture to guess that most people do not even know that the Salvation Army is a Church. The Young Men's Christian Association also began in England and made its way to the US in 1851. The approach of the YMCA is holistic in that the whole man was to be treated body, soul and spirit. Once again, if you were to walk into a typical YMCA it is unlikely that the person at the desk could tell you how to get to heaven. There have been many soup kitchens, missions, homeless shelters and medical clinics that have popped up over the years with the same ambition to correct social ills. The degree of spirituality and evangelism is dependent on the organizers and staff people. However, the overall lack of the gospel and/or biblical training and education has given the "social gospel" a bad name. Even though it is evangelicals who started most of these programs and who are generally the largest supporters of them to this day, still there is a degree of recoil from the idea of getting involved in social causes.

This leaves us with the question as to how should we think biblically about social causes and the gospel? Is there a place for these two to come together and both be effective? Does the church have any obligation to engage in social causes? Does the church have the resources of time, money and energy to engage in social causes?

The answer to any question and particularly these questions, is found in the Bible. So we must first establish how we think biblically about social causes and the gospel. In thinking biblically we will want to lay down a basic premise. The Gospel is good news and is defined in I Corinthians 15:3-4 that Christ died for our sins according to the Scriptures, was buried and rose again the third day according to the Scriptures. This is the basic proclamation of the Gospel. At the same time we understand that there is a bit more to it as well. In the first verse of the Gospel of Mark we read, "The beginning of the gospel of Jesus Christ, the Son of God." A more literal reading would have this as the gospel concerning Jesus Christ, Son of God. This simply means that He not only speaks the message of the good news, He is the good news. In John chapter four the woman at the well was made to realize that the messiah she was looking for and the answer to the great dilemma of her life was standing right there in front of her in the person of our Lord Jesus Christ. We have a key passage of Scripture in Luke 4:18-19. Our Lord was in the synagogue at Nazareth where He was handed the Scriptures. He turned to Isaiah 61:1-2 and read, "The Spirit of the Lord is upon Me, Because He has anointed Me to preach the gospel to the poor; He has sent Me to heal the brokenhearted, to proclaim liberty to the captives and recovery of sight to the blind, to set at liberty those who are oppressed; to proclaim the acceptable year of the Lord." It is possible to spiritualize this entire text and to contend that the Lord was only speaking to spiritual needs. However, a literal translation of the text will cause us to notice that there are at least social implications included here as well as physical healing. It is interesting to note the place where our Lord stopped reading, because the very next phrase reads, "And the day of vengeance of our God;" which is still future and will be accomplished by Him as well.

Let us take notice of each of these phrases which were prophesied concerning the Messiah, which are fulfilled in Christ and which serve to give us the purpose for His earthly ministry. The first of these is "**To preach the gospel to the poor**;" as we have already discussed, our Lord not only preached the gospel, He lived the gospel. But it is to the poor, that He will preach the gospel. Now we know that the poor are more apt to receive it and that it is easier for a camel to get through the eye of a needle than for a rich man to enter the kingdom, but is it possible that there is a reason for our attention being pulled to the poor? A professor of sociology once made the statement that the only thing the Bible ever says about the poor is that they will always be with us. Obviously he has not read much of the Bible. With a quick count in a concordance there are at least 200 references to the poor in the Bible. The law had specific instructions for the care of the poor and the Apostle Paul was instructed by the elders at Jerusalem that the Gentile Churches were to remember the poor. There are other instructions given to the Church concerning the poor throughout the epistles. So not only is the gospel to be preached to the poor but they are to be a concern of the Church which brings the Church into a social issue.

Then the Lord was sent to "**heal the broken hearted**". The gospel will bring peace to a troubled soul and will heal the broken hearted. A broken heart is an emotional strain which will bring on other physical problems. The work of the Holy Spirit in the life of a man or woman will not only impact their spiritual well being but their emotional well being which will also have implications for their physical well being.

He was to "**proclaim liberty to the captives**" and "**set at liberty those who are oppressed**". The gospel is that which liberates the soul from the bondage of sin. In John

8:32 we read, "And you shall know the truth, and the truth shall make you free." Our Lord did not go around getting people out of jail or freeing slaves or overthrowing oppressive governments. But His gospel still sets people free on the inside and it affects how prisons are run and people treat other people who may be subject to their will. In other words there are social implications to this liberty provided by the Lord.

He also healed the blind as well as other many assorted ills, "**And recovery of sight to the blind**". This has to do directly with the physical realm and physical needs. So when we look at this passage, we see both the spiritual side of the ministry and the physical. In fact the two cannot be separated, we cannot give the gospel out of a heart of love and not be pulled into the lives and the needs of the people and their physical needs.

The sermon on the mount in Luke 6 and Matthew 5-8 is a great example of spiritual truth which, when applied, has a great deal of social implications. This is where the words of Ronald Sider apply, "Evangelism, even when it does not have a primarily social intention, nevertheless has a social dimension, while social responsibility, even when it does not have a primarily evangelistic intention, nevertheless has an evangelistic dimension."[12] Evangelism will promote social action **because the gospel changes**

---

12 Ronald J. Sider, *One-sided Christianity?* (Grand Rapids: Zondervan Publishing 1993) 166.

people from the inside out and these people will in turn change their society. During the course of discipleship you cannot help but get involved in the disciples life, their problems and how the gospel is going to change them and solve their problems. "Ultimately, the problems of our world are rooted in sinful rebellion against God. When drug abusers or sexually irresponsible sinners are converted, society improves. When oppressors repent of racism and economic injustice, society improves. New persons create better societies."[13]

This brings me to the subject of what it means to be **a church of irresistible influence** or to put it within this context how should we think biblically about our role as a church in society. The course of the church in modern times has been to withdraw from society and to not seek any role of influence in society as a whole or in the community surrounding the local church. The result is that a chasm has developed between the community and the church that is most pronounced in what is now known as the post-modern world. Robert Lewis, in his book entitled *The Church of Irresistible Influence14* wrote of this chasm and how the church needs to build bridges across the chasm over to our community. It is from this

---

13   Ibid. 177.
14   Robert Lewis, *The Church of Irresistible Influence* (Grand Rapids: Zondervan Publishing 2001)

book that I have gotten the term "**church of irresistible influence**".

The biblical inspiration for this theme is from Matthew 5:16, "Let your light so shine before men, that they may see your good works and glorify your Father in heaven". Earlier in this text (Matthew 5:13-16) our Lord spoke of us as being salt and light which leads to the question of what does it mean to be salt and light?

It is important to note that our Lord does not say that we dispense salt and light but that we are salt and light. The first thing that we learn is that the influence of salt and light comes from what we are as to our being. G. Campbell Morgan wrote, "The influence you exert is always the influence of what you are. No man exerts upon other people any influence by what he says to them, save only as what he says is the outcome of what he is in the deepest fact of his being."[15]

Let us think first of salt. He said, "You are the salt of the earth; but if the salt loses its flavor, how shall it be seasoned?" Or it can be said that if the salt loses its saltiness i.e. that essential property that makes it salt then it is of no purpose. Salt is a preservative and aseptic. It can stop the spread of corruption. It cannot give new life but it can stop the corruption from spreading. It is not difficult

---

[15]   G. Campbell Morgan, *The Gospel According to Matthew* (New York: Revell Co. 1929) 46.

to see how this metaphor carries over into our Christian experience. The Holy Spirit is referred to as the restrainer and the one who judges the world of sin. The Holy Spirit not only indwells each of us but works through us to accomplish this goal. For reasons known only to God, we are not His puppets. His work on the earth is to be carried on by us and our willingness to do it. We are to be out in the world letting the grace and truth of God flow out of us. We read in Colossians 4:6, "Let your speech always be with grace, seasoned with salt, that you may know how you ought to answer each one." We are to be gracious but we are also to be salty.

The same is then true of light. It is that we are light and as a result we give off light wherever we go. Light will drive the deeds of evil into the darkness, why do you think that bars are always dark? Could it be that something evil is being proposed? Throughout scripture light is spoken of as righteousness. Our Lord came as the light of the world, in the eternal state there will be no darkness. In the mean time we are lights, therefore we are to let our light shine out in the open where men and women can be convicted by it.

So what does this mean in practical terms? We find the answer in verse 16, ". . . that they may see your good works and glorify your Father in heaven." It is not in our protest or picketing or writing our senator but it is in our good works that we are salt and light. This is why our Lord prayed for us, "I do not pray that You should take them out of the world," John 17:15. As we are doing good works, this will in turn provide us with the opportunities to give an answer of the hope that lies within us. We will speak with both grace and truth (salt), being salt and light will produce the good works which will provide the opportunities for our speaking the gospel.

Again, we look to the Lord for our example. He spoke of Himself as being the light of the world John 8:12. As we have already noted our Lord was known both for His words and for His works. One of the points of His apologetic in John 5 was His works verse 36. So then it follows that as we follow His example we too will be known for our words and our works. These are the works that will shine as lights so that when the world around us sees these good works they will glorify God. When that happens we then do the best work of all, we give the gospel.

Doing good works is the purpose for which we were saved, Ephesians 2:10 " For we are His workmanship, created in Christ Jesus for **good works**, which God prepared beforehand that we should walk in them." The word *walk* can be seen as the order of our conduct and this is to be done in the sphere of good works. These good works were prepared by God before our conversion so that our salvation

99

is with a view to these good works. Or in the words of Vincent, "God prearranged a sphere of moral action for us to walk in. Not only are works the necessary outcome of faith, but the character and direction of the works are made ready by God." To emphasize this further we see this same point in Titus 3:8, "This is a faithful saying, and these things I want you to affirm constantly, that those who have believed in God should be careful to maintain **good works**. These things are good and profitable to men."

Our life in Christ and His life in us was never intended to be private but to be shared that the riches of His grace might overflow out to others around us. In the same manner the local church was never intended to be a private club where only certain kinds of people are welcome. One of the things that non-believers are constantly trying to convince us is that religion is personal and private. They do not understand that good news is to be shared particularly when we understand what is at stake i.e. heaven or hell.

The next step is for these good works to be done as a local church. So this idea of irresistible influence "is about the great need that exists today of reconnecting the church with the community in a way that makes the church both *real* and *reachable*".[16] It is where the light

---

[16] Robert Lewis, *The Church of Irresistible Influence* (Grand Rapids: Zondervan Publishing 2001) 14.

of Matthew 5:16 and the good works of Ephesians 2:10 give such a powerful testimony of the love of God that people are drawn to the church and to the gospel rather than repelled. Lewis calls upon the readers to use their imagination and to imagine:

- . . . the community in which you live being genuinely thankful for your church?
- . . . city leaders valuing your church's friendship and participation in the community—even asking for it?
- . . . the neighborhoods around your church talking behind your back about "how good it is" to have your church in the area because of the tangible witness you've offered them of God's love?
- . . . a large number of your church members actively engaged in, and passionate about, community service, using their gifts and abilities in ways and at levels they never thought possible?
- . . . the community actually changing (Proverbs 11:11) because of the impact of your church's involvement?
- . . . many in your city, formerly cynical and hostile toward Christianity, actually praising God for your church and the positive contributions your members have made in Jesus name?

- . . . the spiritual harvest that would naturally follow if all this were true?

These bullet points can not only be imagined but can also be set as goals for the church to pursue. This brings us to the question of how does one go about reaching these goals, what strategies would need to be implemented? Rather than use the space here, I strongly recommend buying and reading the book *The Church of Irresistible Influence* by Robert Lewis to get the whole story. It can be purchased at Amazon.com. As with anything we do in the church, it all begins with prayer followed by action. There is an old saying, "Do something, even if it is wrong". The idea being, **do not fall into the trap of the paralysis of analysis**. We need to be willing to try things and see what works and what does not work, but be doing something.

We cannot leave this chapter without admitting that we face serious problems in our society and culture and should at least be aware of them. **First, there is the problem of science and humanism**. The science of the post-modern world is based on the premise that truth can only be arrived at through empirical evidence and logic. This discounts faith and therefore makes Christianity irrelevant in our world. Religion is nothing more than a pacifier for the weak or in the words of Karl Marx "Religion is the sigh of the oppressed creature, the sentiment of a heartless world, and the soul of soulless conditions. It is the opium of the people."[17] Humanist have done an excellent job of getting their agenda well established in our culture with the result, ". . . six out of ten Americans believe the church is irrelevant. And in the lives of the 170 million non-Christians in America (making our country the third largest mission field in the world), that irrelevance provokes an ever-increasing cynicism and hostility."[18] The road back to relevancy will be by way of our good works.

Second, there is the problem of materialism. North American Christians have been so strongly drawn into materialism that we spend most of our money on ourselves. It could also be said that we spend most of our time and energy on ourselves as well. We have become very egocentric. Unfortunately Christians worldwide are not much different. Sider made the following observations taken from research done in 1992:

Christians make up only one third (33%) of the world's people, but we receive about two-thirds (62%) of the world's total income each year. Tragically, we spend about 97% of this vast wealth on ourselves! One percent goes to secular charities. A mere 2% goes to all Christian

---

17 Robert C. Tucker ed., *The Marx-Engels Reader 2ⁿᵈ ed.* (New York: W.W. Norton Co. 1978) 54.

18 Robert Lewis, *The Church of Irresistible Influence* (Grand Rapids: Zondervan Publishing 2001) 23.

work . . . . In 1992, Christians worldwide had a total income of $9,696 billion. We gave only $169 billion (1.74%) to all Christian work.[19]

There does not seem to be any research available which would contradict this study or to suggest that times have changed since then.

Most of the studies completed recently indicate that the average Christian giving is between 2% to 3% of our income. To be honest we would need to include the fact that we are being taxed at a rate of 10% to 15% which will continue to rise as Congress continues to spend at an increasingly alarming rate. The most important question that can be asked on this subject then, is there a place in our budget where we can reduce spending on ourselves and invest it in ministry? The answer to that question will give us an excellent guide as to how much materialism has impacted our lives.

The third problem is one of legalism. Christians are notorious for setting higher priorities on things rather than people. Or to put it in the words of our Lord to Pharisees, "Woe to you, scribes and Pharisees, hypocrites! For you pay tithe of mint and anise and cumin, and have neglected the weightier matters of the law: justice and mercy and faith. These you ought to have done, without leaving the

---

[19] Ronald J. Sider, *One-sided Christianity?* (Grand Rapids: Zondervan Publishing 1993) 191.

others undone. Blind guides, who strain out a gnat and swallow a camel." (Matthew 23:23-24). The result is a much larger proportion of time is spent on making sure everything is doctrinally accurate (in our opinion) than is spent on reaching souls and changing our world. Far too much emphasis is being placed on being sure every t is crossed and I is dotted than on making disciples of the nations. This does not mean that we disregard sound doctrine, to the contrary we should be diligent in teaching the Scriptures, but it is to be sure that we are teaching the Scriptures and not the traditions or opinions of men (see Mark 7:6-9). "All too well you reject the commandment of God, that you may keep your tradition." Mark 7:9. The result is that while we are arguing over things that ultimately do not matter, souls are plunging into hell and unbelievers are driven away by our legalism and the resulting hypocrisy.

So where does that leave us? It ought to be clear that the Bible addresses more than spiritual issues. The healing ministry of our Lord confirmed who He is and it demonstrated His concern for the physical ailments of the people around Him. He ate and drank with sinners and reached out to the poor. His disciples followed His example as we see in the book of Acts. Even more than that, the teachings of Scripture (our dogma) will invariably have social implications. The plain fact of the matter is that Believers, living according to the Bible, will make a better society than nonbelievers. It is also clear that we were saved with a view to ordering our conduct within the sphere of good works. The purpose of these good works is to draw men and women to our Lord. Therefore, our ultimate good work or the goal of any ministry is to give the gospel. **To do ministry without the gospel is to perform yet another social program which in the long run will do nothing to change society**. To do ministry with the gospel provides the avenue by which a person can be changed from the inside out and then move to change the world around them. This kind of change has been proven to work, not only in our culture but in cultures all around the world.

The challenge remaining for us is to become a church of irresistible influence which begins with each individual believer being committed to doing ministry that has as its ultimate goal to preach the gospel. The difficulty is in how we invest our time, money and energy. Our stewardship consists of more than just our money, it includes our time and energy. By energy, I am including how we use our gifts and abilities. We all have 24 hours in a day and 168 hours in a week. If we use 8 hours per day for sleep, which is 56 hours per week and we work 8 hours per day for 5 days and 8 hours per day in whatever on weekends then that takes another 56 hours per week. We are then left with 56 hours to eat, spend time with family and recreation. The question then becomes, is it possible to leave 1 to 2 hours for ministry inside the church, another 1 to 2 hours for ministry outside the church and another 1 to 2 hours for serious study of the Scriptures for a total of 4 to 6 hours per week? This will leave another 50 hours for other things. Keep in mind that ministry can include the family and this kind of family time is some of the most valuable. This includes the energy, gifts and abilities to perform the task which are taken up by the time being spent. So when we think of time, we also have to think of the energy expended during that time.

Our Lord taught in Luke 12:22-34 a lesson in priorities. First, do not be concerned about these temporal things as the Lord will provide. Then secondly, invest in the things of heaven, "where no thief approaches nor moth destroys" (verse 33) because where your treasure is that is where your heart will be as well (verse 34). I have often thought that the reverse is also true, that where your heart is, that is where you will invest your treasure and where you invest your treasure will hold your heart. When we seek first the kingdom of God, then that is where we invest our treasure and that will hold our heart.

In the sales profession there are 3 questions that need to be answered in order to make a sale; can they afford it? Do they need it? Do they want it? Every sales person knows that if a person wants something bad enough, they will find a way to afford it, justify that they need it and they will buy it. Therefore the emphasis is always on the "want it" factor. The question to be answered is how badly do we as individuals and then we as a church "want it"? How badly do we want to see souls come to know the Lord and our society influenced to the glory of our Lord?

Read more about it

A Theology As Big As The City by Ray Bakke
One-Sided Christianity? by Ronald J. Sider
The Church of Irresistible Influence by Robert Lewis
The Gospel According to Matthew by G. Campbell Morgan

CHAPTER SIX

EVANGELISM—WHY BOTHER

It was a hot July in Omaha with daily temperatures near the 100 degree mark. A youth group from a small town in Western Nebraska had come for their mission's trip to North Omaha. I was pleased and excited to provide this opportunity for these bright and enthusiastic young people and provided them with the opportunity to do a Bible club in an old, historic church/school in the North Omaha community. The church building did not have air conditioning and it was very hot and humid. By the time the week was over, I was totally drained and kept thinking that I was getting too old for this stuff. I had just given the final invitation for the children to trust Christ on the last day when a little girl came forward and pulled on my pant leg to get my attention. I kneeled down to her level, she was dressed in a beautiful white dress which highlighted her black skin. She then spoke to me and said, "Uncle Greg, today I placed my trust in the Lord Jesus". She said the words just as I had been teaching them all week.

Greg Koehn Ph.D.

I leaned over and she placed her small arms around my neck and I took her into my arms, thinking, this is why I am here and this is why it is worth it. Evangelism is never about programs, presentations or things, it is always about people. So what motivates you to give someone the gospel?

Having discussed our theological basis (Soteriology), the Psychology and Sociology of Evangelism it is now time to get down to the nitty gritty. What does "Evangelism" mean exactly and what is the "Gospel" that we are to give to people and why should we bother?

The words gospel, good news and evangelist all come from the same root word uangelion (euaggelion) which is translated gospel. The word uangelizo (euaggelizw) is the verb form which is to preach the gospel or good news. The word uangelistees (euaggelisths) is an evangelist and is where we get the concept of evangelism and evangelical. People often wrestle with the word "evangelical" because of the way in which the word is abused in our culture. By definition an evangelical is simply one who believes in giving good news, so why is that a problem? I often ask people if they like good news and would like to hear some? If they say yes, I give them the gospel. You will notice that in the middle of the word is angel, an angel is one who brings good news such as in Luke 2:10 and the announcement of our Lord's birth.

To help us understand the role of an evangelist let's look at some evangelists in the Scriptures. The greatest example of all things is our Lord. As was mentioned in a previous chapter our Lord is seen not only as preaching the gospel but as the gospel personified. A more literal rendering of Mark 1:1 can be, ". . . the gospel concerning Jesus Christ". This is a key point in that the gospel is simply the introduction of Jesus Christ as Lord to a lost and dying world that is badly in need of His resurrection power and life. When our Lord preached the gospel as in Mark 1:14,15 He was presenting Himself as the King of the Kingdom of God (Kingdom of Heaven in Matthew). This can be seen in every instance where our Lord was speaking either to individuals or to groups that the direction of focus was to Himself. He is the benevolent king who provides for His subjects, brings order and justice to His realm and will protect from all enemies. He provides the abundant life now and eternal life later. Our past, present, and future suddenly find their fulfillment by faith in Him.

Our Lord sent His disciples out to preach the gospel of the kingdom, "And as you go, preach, saying, The kingdom of heaven is at hand." Matthew 10:7. There are at least four things our Lord gave them as He sent them out. First, they went with the power to do miracles esp. heal the sick, cleanse the lepers, raise the dead, and cast out demons. This message of the kingdom had to do with both the physical and the spiritual. Second, they would live off the generosity of those hearing the message. They did not provide any money for their journey, neither a second set of clothing because the workman is worthy of his food. If there was no one in the city to receive them, then they would leave that town to its own destruction. Third, they could expect to be persecuted and rejected. They would go out as sheep among wolves and the wolves would turn on them. They will be like their teacher and as their teacher was rejected so would they. At the same time they could have courage in their message as it would be given them what to say when they needed it. Fourth, where the price of discipleship may be great but the reward is greater. People can only kill the body, they cannot destroy the soul and those who lose their life here will gain it in eternity (Matthew 10:5-42).

The problem with our Lord as an example is that He is God and with that comes certain advantages. Fortunately the Bible has other examples of men who are basically just like us. Peter is an interesting character in that he denied the Lord 3 times. It seems like he was always putting his foot in his mouth. However, we know that God greatly used him to advance the cause of the gospel. We have a great example of what it meant for Peter to serve the Lord in the gospel in Acts 5. Peter is arrested with other apostles and put into prison, but an angel comes and sets them free. Peter is told to then go and stand in the temple and speak the words of this life, verse 20. Note the words to **go, stand and speak**, a great summary of our task in evangelism. Peter did that but is arrested again and ordered to not speak in the name of Jesus. His answer is very instructive for us to this day, "We **ought to obey God rather than men**" (verse 29). Here the command of God to preach the gospel takes precedence over the command of the civil/religious authority. When the counsel became furious over this answer, Gamaliel had the apostles step outside while he addressed the assembly and pointed out that if this is of man it will come to nothing, but if this is of God, nothing can be done to stop it (verses 38-39). This is an excellent point to keep in mind, that which we do in the preaching of the gospel is of God and ultimately there is nothing that can be done to stop it. So let us insist on being on the winning side. There are hazards in route to our final victory as Peter and the apostles were taken and beaten. Their response was to rejoice in that they were counted worthy to suffer for the cause of Christ (verse 41). It is easier to suffer for Christ when you have truly appropriated the ultimate victory. Peter then went right back to preaching daily in the temple (verse 42).

In Acts 8 we see Philip preaching in Samaria just as was told them in chapter 1:8 that the gospel would spread from Jerusalem to Judea and to Samaria. In fact we read in verse 4 that they were scattered everywhere preaching the gospel.

Then we have the Apostle Paul who was set apart for the preaching of the gospel (Romans 1:1) to the Gentiles specifically. As he wrote, by inspiration of the Holy Spirit, to the Church at Rome, he wrote of his eagerness to preach the gospel in Rome (verse 15). He went on to describe the gospel as the **power** (dunamis, δυναμις) of God **unto salvation** (verse 16). This word *power* is a word which means that there is power inherent in its nature. The gospel contains within itself that power to give salvation to everyone who believes. It is like holding dynamite in your hands and wherever it explodes it changes lives so that they will never be the same again.

So then the importance of the gospel can be seen in those who first delivered it. This importance can also be seen in the fact that God has chosen to carry out His sovereign will through evangelism. It is important to keep in mind that **salvation is of the Lord (Jonah 2:9)**, God the Father chose us in Christ before the foundation of the world. We were predestined to adoption as sons by Jesus Christ. The word *predestined* is to set boundaries around beforehand. What these boundaries are is never specified in Scriptures, but we do know certain things that are true. We know that being born in a Christian family is a greater advantage toward trusting Christ that being born in a non-Christian family. That being born in a country where the gospel is readily available and where there is freedom of religion is an advantage. We know that being well fed, clothed, housed and educated is an advantage. All these advantages and more can add up to the boundaries that have been set around us so that there is a point at which we trust Christ. Obviously God can, in His sovereign will, overrule any disadvantages, an evangelist can make his way to the person who lives in a country that is poor, undeveloped and very little opportunity to hear the gospel and give the gospel to a person who then believes. Even this is within the boundaries set by God before the foundation of the world. Whatever else may be the situation we know one thing for sure and that is that God carries out His divine will for salvation by sending a messenger with the message of the gospel, the person who has been made alive by the Holy Spirit will hear that message, understand the message and believe the message to the point of believing in the Lord Jesus Christ. We have an excellent example of this in **Acts 13** where Paul is preaching first to the Jews and then to the Gentiles. The Jews rejected the message and Paul turned to the Gentiles (verse 46) then we read that upon hearing the message ". . . as many as had been appointed to eternal life believed." (verse 48). Again, in **Acts 16** we read of Paul going to the city of Philippi and met Lydia, "The Lord opened her heart to heed the things spoken by Paul." **(verse 14)**. The essence of evangelism is to give the opportunity to hear the gospel to everyone, all the while looking for a divine encounter where the Holy Spirit has opened their heart to understand it and believe it (Ephesians 2:4-5).

So now we see these two components of salvation working themselves out according to the sovereign will of God. First we have been chosen in Christ before the foundation of the world, predestined to the adoption as Sons all by the good pleasure of the will of God **Ephesians 1:4-5**. There are none who seek after God **Romans 3:10,11** and we are all dead in trespasses and sins until the Holy Spirit makes us alive **Ephesians 2:5.** Second, we still have the responsibility to believe **John 3:18; 5:24**. The gospel is to be preached and it is to be preached that *whoever calls on the name of the Lord* **Romans 10:13**; that God loves the whole world **John 3:16** and that God is not willing that any should perish **II Peter 3:9; I Timothy 2:4.** If you have trouble reconciling these two components, do not worry about it because there are certain things about God that we will never understand because God is far above us and His ways are not our ways. If we knew all there was to know about God, He would no longer be God but the product of our imagination.

Some people struggle with the sovereignty of God in salvation because it goes back to our motivation for evangelism. If God chose us before the foundation of the world then why should we bother to evangelize? This was the position of the Puritans long ago that salvation is something that is worked out, if you were the elect then it would work out, if not then it would not. In reality those who believe in the sovereignty of God in salvation are more apt to preach the gospel than those who believe that it is all up to the "free will" of man. As evangelist we are not sales people seeking to persuade people with our eloquence that this is the best decision of their life, after all what kind of moron would chose to go to hell rather than heaven? The problem is that the people we are talking to are spiritually dead and dead people do not make decisions. Rather we are those who are obeying the direct command of God to make disciples of those who the Holy Spirit has made alive to listen, understand and believe the gospel.

As noted in chapter two the sovereignty of God also guarantees our salvation **Ephesians 1:13-14**. Inasmuch as salvation is of the Lord it cannot be lost. Salvation is dependent upon what God has done for us rather than on what we have done for God **Romans 8:28-39**.

Some will say that the sovereignty of God in salvation is unfair, as it smells of determinism, that people go to hell because God did not choose them to salvation. It is important to remember our basic soteriology that we are sinners from the womb and therefore deserve to go to hell. Then during the course of our life time we earn the right to go to hell. The question goes to the question of **Romans 9:14, "Is there unrighteousness with God?"** The answer is found in the remainder of the chapter where we see that as a sovereign God He may do whatever He wills, to whomever He wills, whenever He wills. He is as a potter who has power over the clay to make one lump to honor and another to dishonor. May the lump of clay protest to his creator, why did you make me this way? The answer is no, we are not in a position to contend with God as to doing what He does according to the good pleasure of His will.

So what is this gospel that we are to be preaching? We have noted previously that the gospel is good news, great, can we be more specific? Once again let us allow Scriptures to speak to the issue at hand. In **I Corinthians 15:1-8** we have this gospel that we are to preach defined for us in plain and simple terms. In verse 1 we have the gospel that was preached. It is important to keep in mind that the gospel must be preached, there is a point in time when this message must be verbalized. Nobody **is going to heaven because they noticed that Christians are such sweet hearts**. Our life style may provide opportunities for giving the gospel but we must tell the story in order for people to have the opportunity to hear, understand and believe. Then we notice how those in Corinth heard the gospel, then they received it as being true and then they made their stand in the gospel i.e. they believed it.

It is important to notice the steps one will take to trusting Christ. First there has to be knowledge of the facts of the gospel. This knowledge comes by way of a preacher who comes bearing good news **Romans 10:14-17.** Then there is an understanding of the gospel, the Corinthians received the message i.e. they received it as being true. There are those who contend that believing the gospel is true is sufficient for saving faith. The simple fact is that there are millions of people who believe the gospel is true but have chosen to reject it as not being for them for a number of reasons. Even the demons believe and trembles (James 2:19; Matthew 8:29). We see this process in **John 1:12,** we first read in verse 11 that He came to His own and His own did not receive Him. At this point it looks like His entire ministry has failed. But they we read in verse 12 that as many as do receive Him, in other words there will be those who do receive or embrace Him as being true. Then the verse continues with ". . . to those who believe in His name:" The third step of the process is to believe in His name. In the same manner the Corinthians took their stand in the gospel, this is where they placed their faith. Believing faith then is more than knowledge or an agreement to the facts but it is a commitment to those facts.

The gospel, as defined in I Corinthians 15,consists of 4 points. The first is that **Christ died for our sins according to the Scriptures** verse 3. We may very well ask which Scriptures? The answer is not hard to find as we read of the suffering servant in Isaiah 53 and especially verse 5, "But He was wounded for our transgressions, He was bruised for our iniquities; the chastisement for our peace was upon Him, And by His stripes we are healed." The second point is that **He was buried**. This is important because it confirms the fact that He died. You only bury dead people and the Romans who witnessed His death know what a dead person looks like (they killed enough people) and the women who prepared Him for His burial knew what they were doing and knew He was dead. Third, **He rose again the third day according to the Scriptures.** Scriptures referred to here include Psalms 16:10, "For You will not leave my soul in Sheol, Nor will You allow Your Holy One to see corruption." The fourth point verifies that He rose from the dead in that He was seen. In a court of law the strongest argument for a case is in the eye witness account. The eye witness account of our Lord was that He was seen first by Cephas, then by the twelve, then by over 500 brethren at once. After He was seen by James and the Apostles, He was seen by the Apostle Paul himself. There are two things that are well to keep in mind in considering the eye witness accounts of the risen Lord. The first is that they are credible witnesses. From the women who came first to the tomb, to the disciples and then to over 500 people all were seen to be honest people who were not inclined to make up fabrications about something so important. Second, as I Corinthians 15 goes on to explain the resurrection lies at the core of our salvation, "And if Christ is not risen, your faith is futile; you are still in your sins" (verse 17). If the resurrection were not true, then out of the hundreds of people who confessed to have seen Him, just one could have spoke up and stopped the whole Christian movement before it ever got started. Many of these people, if not all, were later tortured for their faith and martyred. To stop the torture all they would have had to do was deny the resurrection and they would not, simply because they would not speak against the truth. Frank Morison wrote a book which began as an attempt to disprove the resurrection from an historical basis using the same kind of investigative techniques and arguments a lawyer would use to present a legal brief. He ended up writing a much different kind of book.

It is not that the facts themselves altered, for they are recorded imperishably in the monuments and in the pages of human history. But the interpretation to be put upon the facts underwent a change. Somehow the perspective shifted—not suddenly, as in a flash of insight or inspiration, but slowly, almost imperceptible, by the very stubbornness of the facts themselves.[20]

Others have researched and attempted to disprove the resurrection but in every case where there is an honest evaluation of actual historical evidence the result comes out the same that on the third day Jesus rose from the dead. This subject will be examined further in our chapter on Apologetics.

There are those who believe that **a necessary component to the gospel is repentance**. It is important to keep in mind what the word "repent" (metanoein) actually means a change of mind or of direction. In order for a person to stop and listen to the gospel they will have to change their mind about what they have thought of God in the past. It is interesting to note that the word "repent" does not appear in the Gospel of John. In the majority of instances in the New Testament it is used of Israel's need for repentance in preparation for their Messiah. The Jewish audience well understood what it meant to repent as the prophets had been calling on them to repent for hundreds of years. The ministry of John the

20 Frank Morison, *Who Moved The Stone?* (Downers Grove: InterVarsity Press, 1971) preface

Baptist was one of repentance to prepare the way of the Lord (John 1:23). It is true that there are two places in the book of Acts where it is obvious that the message of repentance is being given to the Gentiles in Acts 17:30 "Truly, these times of ignorance God over looked, but now commands all men everywhere to repent," and Acts 26:20 "but declared first to those in Damascus and in Jerusalem, and throughout all the region of Judea, and then to the Gentiles, that they should repent, turn to God, and do works befitting repentance." When preaching the Gospel to a whole crowd of people or in addressing representatives of our or any country, the call for repentance is certainly appropriate as the nations of the world need to change their mind and the direction of their people with regard to God, His Word and His salvation found in Christ. When working with an individual, you may find it difficult to insert the word repent or to define what that means prior to salvation. For example in Acts 16:31 when Paul was asked by the Jailer what must be done to be saved, his reply was simply "Believe on the Lord Jesus Christ and you will be saved . . ." Most often that is what the sinner needs to hear as they have already changed their mind about what they think about God and their relationship to Him. But if you feel compelled to insert the word *repent* in your presentation of the Gospel, feel free to do so, but be prepared to explain what exactly you mean. A word of caution here, do not use repentance as a tool to insert your own ideas of holiness and to require people to do works that are impossible for them until they have received the Holy Spirit. For example, a person may not be able to give up drinking alcohol until after they have trusted Christ and are born again.

Now that we have an understanding of the simple message of the gospel all we need is the motivation to give the "good news" to others. At this point we will examine 4 motivators for evangelism. There are more than that but for our purposes we will look at these four, only.

The **first** of these is **how we view authority Matthew 28:18-20.** When our Lord gave what is now known as the "great commission" He began with, "All authority has been given to Me in heaven and on earth." The command in this text is not to go but rather as we are going to make disciples. The word *go* is from πορευθεῶτεs which is a participle, simply meaning that it describes both the action and the person. The one who obeys the Lord will be one who is going into the world where he will make disciples by baptizing and teaching them. Whether or not we obey the command to make disciples of the nations will depend on how we view authority. While I was growing up, my father would give a command and we did it without question or hesitation because he had taught us to respect authority to the point where obedience was instant. However, as we know, not everyone views authority the same way. Therefore, there will be those who will see these words as a part of holy inspired scripture but not practically significant in their daily lives.

The **second** of these motivators is found in **II Corinthians 5:14** which is **the love of Christ,** "For the love of Christ compels us, because we judge thus: that if One died for all, then all died;". The word *compels* can also be to be constrained or held fast. We can picture this as a rope of love that wraps around us so that we must obey out of love. We show our love for God when we keep His commandments (I John 5:2-3). The essence of salvation is found in the love of God which is why we are called upon to "Behold what manner of love the Father has bestowed upon us, that we should be called the children of God!" (I John 3:1). God so loved the world that He gave His Son (John 3:16). When our Lord walked upon this earth, He was moved with compassion when He saw the multitudes (Mark 6:34). In following our Lord's example and because of His love in our hearts we can be motivated to give the gospel because of this love. From John 3:16 we can learn that the greatest possible gift of love that can be given is to give someone the gospel. This far exceeds any material, physical thing that we can do. So when someone says that you must first prove to people that you care, the giving of the gospel, in and of itself proves that we care to the point of love.

The **third** of these motivators is found in **II Corinthians 5:11 the terror of the Lord.** We know about the fear of the Lord that is an awesome respect. But here we have the terror of the Lord which is to understand that there is coming a time in which we will give an account of our lives before the Lord and at that time it will be too late to do things better or differently. So here we have the Apostle Paul (by inspiration) looking ahead to the judgment seat of Christ (Vs 10) and knowing that at that time he must give an account of what he is doing now, therefore he will persuade men. This is often illustrated by the story of a boy who lived on a farm. One morning his father gave him instructions to weed the garden. The boy knew his father would not be home until sunset so he had plenty of time. Unfortunately he procrastinated the whole day and as the sun was going down, he was feverishly working in the garden because he knew if he did not get the job done by the time his father returned home it would not go well with him. To do what our father has told us to do can be out of an awesome respect for our father (fear) but to not do it and then have to give an account for it is terror. Therefore, knowing that this day of giving an account is coming we, like the Apostle Paul, need to be persuading men with the gospel.

The **fourth** of these motivators is for **reward** as seen in **I Corinthians 3:13 and II Corinthians 5:10.** In I Corinthians 3 we have the teaching that we are to build on the foundation of our Lord. How we build will then be evaluated at the judgment seat of Christ as passing through fire and what was done poorly will burn off as wood, hay and stubble, but what was done well will come through the fire as gold, silver, precious stones. This day will declare our works of what sort they are and those works which come forth as the gold, silver and precious stones will result in a reward. In the days of my youth I ran track in High School. As a result I received medals for my efforts. Years later when my son was looking through my scrap book, he came across these medals and ribbons. I noticed that what was once gold had tarnished and turned green. I had worked so very hard at what turned out to be a perishable crown. It is natural to be motivated by the chance for a reward, let it be this imperishable crown of eternal glory in leading others to this great salvation.

So this brings us back to the beginning question, what motivates you? Our message of good news is not difficult to explain, we are simply looking for divine encounters which the Lord has already prepared for us, all that is left for us is to overcome our fear with boldness and do it!

CHAPTER SEVEN

IN DEFENSE OF THE FAITH

One of the most, if not the most common objection to the Gospel is "I don't believe there is a God". Upon one occasion, after a lengthy discussion with an atheist, I pointed out that the Bible mentions him, "The fool has said in his heart there is no God" (Psalm 14:1). This, of course, ended the conversation. In this day and age known as the postmodern world, more and more people are being a fool and saying that there is no God. This is known as atheism, which comes from the Greek αθεоs. The word for God is Theos (θεоs), place an alpha in front of a word and that negates it so that in this case God becomes no God. This is known as the alpha position. This is a position that is very difficult to defend, you simply ask, how do you define this God that you do not believe exists? If they even begin to answer then they have acknowledged that the <u>possibility</u> of God exist because you cannot define something that does not exist. This is when an atheist becomes an agnostic which simply means "do not know". The problem with an agnostic is not God but a lack of knowledge or ignorance.

When we seek to defend the existence of God, the validity of Scripture, the deity of Christ etc., this is known as apologetics. It is important to keep in mind that apologetics consist of a theological, philosophical and logical defense of the truthfulness of Christianity.[21]The word for apologetic comes from the Greek απολογια (apologia). which literally means "from logic". The Scriptures instruct us to use apologetics in I Peter 3:15 ". . . and always be ready to give a *defense* . . ." and throughout the Acts (see Acts 17:22-34) and the Epistles we see a logical defense of the Faith.

Now back to the logical argument as to the existence of God. There are at least five main philosophical arguments for the defense of the existence of God. They are the Cosmological, the Teleological, the Ontological, the Anthropological and the moral arguments.

Cosmological: This comes from the word cosmos, which means order. The key to this argument is based on the physical law of cause and effect i.e. for every effect there is cause and the cause is greater than the effect. For example should a rock fly through a window who would ever think that the rock picked itself up and threw

---

21   Dr. Jim Anderson, *Christian Apologetics A Defense Of The Faith* (Springfield MO: Video bible Studies International, 1990), xi

itself through the window? That would be nonsense. So how much more nonsense is it, that a very immense and complex universe created itself? There is another aspect to this argument known as infinite regress. This simply means that at the beginning there must be an uncaused cause. In other words, the question of "what caused that?" must stop with the ultimate cause that itself was not caused. Aristotle called that the unmoved mover. We call Him God!

Dr. Jim Anderson points out three main points where we can see cause and effect. The first of these points is **Immensity.** Space and the number of stars present a universe that is too immense to even behold, let alone measure and categorize. The Scriptures tell us that to behold the heavens is see that there is a God Psalms 19:1; Romans 1:20. As scientist have invented and used larger and larger telescopes they have come to the conclusion that the stars are too many to be counted. "The total number of stars in the observable universe is estimated to be 1025 (1 followed by 25zeroes). Nobody knows the actual number." [22] A survey was taken which showed that 90% of all astronomers today believe there is a God. Of course they do not publicly acknowledge such a thing because they might lose their tenure at the universities

---

[22] Werner Gitt, *Counting the* Stars; [article online]; available from http://www.answersingenesis.org/creation/v19/i2/stars.asp; internet.

where they teach.[23] Such a universe could not have happen by chance and must have been created by one who is even more immense than the universe. The second of these points is **Complexity**. When we look at the universe from a simple leaf to the human body we see complexity which demonstrates an infinite intelligence.

> Quantum physics has demonstrated that at the level of subatomic particles, there is an irresistible urge of electrons toward symmetry and that there is an amazing cosmetic aspect to the universe. One author said that nature is a great architect, meaning that nature is God. It is also a great astronomer, a great chemist, a great physiologist, a great psychologist, and a great mathematician, demonstrating an incredible knowledge of the facts of the various sciences now known to mankind, which have all said the same thing.[24]

The third of the points is **Undeniability.** The universe with its immensity and complexity is real and cannot be denied. According to the law of cause and effect, the ultimate cause of the universe must be real as well. The universe is screaming at mankind that there is a God who is the uncaused cause of everything we see and all that we are in ourselves.

---

23 D. James Kennedy, *Why I Believe* (Dallas: Word Publishing, 1980) 39.

24 Ibid., 40-41.

Teleological: This comes from the word teleos, which means design. This argument is closely related to the previous in that it also has to do with creation. True science is based upon observation followed by a theory, followed by testing, followed by a hypothesis, followed by more testing which will come to either a hypothesis or null-hypothesis. By observation we can see that our universe is very complex. We also know, by observation that complexity cannot come from chaos or order cannot come from disorder. Order or complexity can only come from order or a greater complexity. To say that our universe came from chaos is counter to the laws of natural science coming from our observation of the natural order of our world. Again, the greater complexity or the grand order, which gave order and design to our world, is God!

Of all God's creation the one part that speaks to the existence of God more than any other is Man. "So God created man in His own image; in the image of God He created him; male and female He created them." (Genesis 1:27). When God chose to communicate Himself to mankind, in terms that we could understand, He took upon Himself humanity i.e. He became man (John 1:14; Philippians 2:7). Therefore, of all of God's creation the one part that best speaks to the existence of God and communicates God is Man. Is it not interesting that when we want to get away to experience God, we leave town and go somewhere out in the woods, mountains etc. Maybe it is because we have lost the implications of what it means to be created in the image of God and are in desperate need to get that back. Which then leads us to the presentation of the Gospel. This then takes us back to the list of arguments we use in apologetics to defend our Faith.

Anthropologetical: This is the argument from man i.e. anthropos is man in Greek and anthropology is the study of man. When we study man, (man is used as the generic term for humanity, made up of both men and women) we see that there are certain specific differences between man and animals. There are many physical and emotional differences between the beast and us but for the sake of our argument here we will focus on three, which are in the realm of the soul and spirit. It should be noted that man was created in three parts, consisting of Spirit, Soul and Body (I Thessalonians 5:23). It cannot be proved that animals have anything more than bodies and a form of an intellect that is trained by environment, circumstances etc. Therefore, the three that we will focus on are *intelligence, moral discernment and person-hood*. It is obvious that man is vastly more intelligent than animals; even people who behave as though they are only one notch above a monkey are in fact far more intelligent than beast. It has been said that you could place 100 monkeys in front of 100 typewriters for 100 years and they still would not be able to write one line of Shakespeare. Man is far superior in intelligence, which is why God put him in charge of all creation. The second is moral discernment, which is why man organizes society and has laws in place for the greater good. Man is the only one of God's creation who has a concept of right and wrong and creates society accordingly. The third is person-hood, which includes everything that makes up a person such as the ability to reason, to have emotion and to be creative. All three of these point to a creator who created man in His image.

Ontological: This comes from the Greek word Ontos which simply means "being". This argument connects to the anthropological argument in that the idea of God pervades all mankind. That God is a being and has placed within man the idea of God. The Scriptures refer to the "light that lights every man" (John 1:9) and the conscience that bears witness (Romans 2:15). As a result mankind is deeply religious. The communist and other atheist try to stamp out religion but cannot do it. An idea has to come from something real, therefore this idea of God must come from God.

The Ontological argument was further developed by Anselm, archbishop of Canterbury (A.D. 1033-1109) to what has become known as "perfect being theology." This is summarized in his statement "God is a being with the greatest possible array of compossible great-making properties."[25] "A *great-making property* is any property, or attribute, or characteristic, or quality which it is intrinsically good to have, any property which endows its bearer with some measure of value, or greatness, or metaphysical stature, regardless of external circumstances."[26] Simply stated, this means that all the great things that we think of as attributes of God that are not contradictory, all come together in this one being which is by definition God. An atheist may well respond by saying that this will define God but does not prove His existence. That then begs the question, then where did the idea of all these great attributes come from? Therefore, if God does not exist then neither do these great making attributes such as intrinsic, unconditional love; an all powerful source of energy and matter, which is the uncaused cause; an absolute standard of what is right and good etc.

Other philosophers such as Norman Malcolm developed the argument further. The simplest way of stating the argument is in a logical format:

1. If God exists, his existence is necessary
2. If God does not exist, his existence is impossible
3. Either God exists or he does not exist.

---

25 Thomas V. Morris, *Our Idea of God* (Notre Dame: Notre Dame Press, 1991) 35.
26 Ibid.

4.   God's existence is either necessary or impossible.

5.   God's existence is possible (not impossible)

6.   Therefore God's existence is necessary.[27]

Therefore, if a person is willing to admit that such things as an uncaused cause, an infinite source of power, intrinsic love and an absolute standard of what is true and right then one, logically must admit that at least there is a possibility that God exists. With the possibility of God's existence comes the logical conclusion that either God must exists or must not exist, therefore He exists.

Moral: Next is the moral argument. Man is moral in that all of mankind has a sense of right and wrong. Morality is not something that can evolve from lower life forms as none of them have a morality. Morality and the sense of right and wrong can only come from an ultimate law giver or the absolute right. Again, this is God.

This argument that morality is built into humans is developed in the Bible in the book of Romans beginning in chapter one where we read, "For the wrath of god is revealed from heaven against all ungodliness and unrighteousness of men, who suppress the truth in unrighteousness, because what may be know of God is manifest in them, for God has shown it to them."

---

[27]   C. Stephen Evans, *Philosophy of Religion* (Downers Grove: Intervarsity Press, 1982) 48.

(1:18-19). We see that the wrath of God is coming upon those who suppress the truth. How could they suppress the truth unless they know what it is? The answer is found in this next statement of judgment which is the fact that God has showed man what is true and right but man has ignored that revelation. This argument is advanced further in chapter two where we read, "for when gentiles, who do not have the law, by nature do the things in the law, these, although not having the law, are a law to themselves, who show the work of the law written in their hearts, their conscience also bearing witness, and between themselves their thoughts accusing or else excusing them." (2:14-15). Everyone is familiar with the concept of the "conscience" that which is on the inside of all of us, telling us what is right and wrong. Our conscience is the law written on our hearts and it was given to us by God when we were created in the image of God.

Here is the core of the argument as there are those who argue that our understanding of what is right or wrong is depended upon social evolution where over the course of time mankind gradually figure out what is best for their culture and society and then behave accordingly. Social evolution has the same problem as physical evolution with the law of cause and effect. There has to be a point in which there is the first standard of right and wrong from which all others follow because there is no such thing as infinite regress.

In order to prove this, we need only look at the history of societies and see if there is a common thread of morality and/or law which would lend itself to a common source. That common thread is what is known as common or natural law. Historically one of the foremost minds on the subject is a man by the name of William Blackstone. He collected the miscellaneous writings on common law and put them all together in a volume known as the Commentaries published in 1765. His was one of the writings which greatly influenced the thinking of the men who wrote the Declaration of Independence and our Constitution. Note just some of what he wrote:

> LAW OF NATURE—This will of his Maker is called the law of nature. For as God, when He created matter, and endued it with a principle of mobility, established certain rules for the perpetual direction of that motion; so, when He created man, and endued him with free will to conduct himself in all parts of life, He laid down certain immutable laws of human nature, whereby that free will is in some degree regulated and restrained, and gave him also the faculty of reason to discover the purport of those laws.
>
> Considering the creator only a Being of infinite power, He was able unquestionably to have prescribed whatever laws He pleased to His

creature, man, however unjust or severe. But as he is also a Being of infinite wisdom, He has laid down only such laws as were founded in those relations of justice, that existed in the nature of things antecedent to any positive precept. These are the eternal, immutable laws of good and evil, to which the Creator Himself in all his Dispensations conforms; and which he has enabled human reason to discover, so far as they are necessary for the conduct of human actions. Such, among others, are these principles: that we should live honestly, should hurt nobody, and should render to everyone his due; to which three general precepts Justinian has reduced the whole doctrine of law.

This law of nature, being coeval with mankind and dictated by God Himself, is of course superior in obligation to any other. It is binding over all the globe in all countries, and at all times: no human laws are of any validity, if contrary to this; and such of them as are valid derive all their force, and all their authority, mediately or immediately, from this original.

To summarize, it simply means that just as God has built into His creation certain physical laws, He has also built into His creation mankind certain laws that determine right from wrong, good from evil. These laws are seen in nature and in God's revealed law the Bible. For this reason we can study any society with its culture(s) and see that there are certain laws that are consistent with all other societies such as honesty, hurt nobody, and to render to each one his due.

Our country was founded and has enjoyed prosperity because we have continually held to this principle that there are certain unalienable rights that have been given to us by our creator and the law of the land is to protect those rights because they have been given to us by our creator and therefore there are no better rights or better law or better morality than that which was given by our creator. (The use of redundancy in the last sentence is given for the purpose of emphasis).

In his book *Discipling Nations* Darrow Miller writes concerning "Natural Law" and its impact upon a culture:

> Natural law has been acknowledged down through history. Not only Christian writings but also Hindu, Chinese, and Greek literature support this concept. Augustine, Aquinas, and Calvin all wrote about it. The great Christian theorist John Locke influenced the deist Thomas Jefferson to refer to natural law in the Declaration of Independence.[28]

With Natural Law and the absolute morality upon which it is based, then humanity is confronted with the choice of either to do good or to do evil and this choice

---

[28]   Darrow L Miller, *Discipling Nations* (Seattle: YWAM publishers, 1998) 129.

or series of choices have a profound impact upon culture and then upon society as a whole.

> Man was given moral freedom to make choices—significant choices. The secularists believe that man is a machine, an automaton. Animists, for their part, believe that man is dominated by outside forces. The truth is, we have real freedom. This means we face both real choices and genuine consequences. Man is the proactive creator of history, not an inactive fatalist or a reactive responder. In contrast to Hinduism and Buddhism, which hold that "man enters the water and makes no ripples," theism teaches that "man enters the water and makes ripples that go on forever."[29]

These moral arguments are important because they lie at the crux of the matter. I once was talking to a clerk at a Hotel where I was checking in. He made the comment that he did not want to become religious just wanted to know what the Bible said. I replied by asking him the question, "do you believe in infinite regress?" He responded that he knew where I was heading with that question and that he did believe that there is a God. I then pointed my finger at him and told him that I knew what his problem is, that if there is a God then he would have to give an account to that God and he did not want to do

---

[29]   Ibid. 130

that. He agreed and I went on to give him the gospel so that he would be able to face the God who is righteous and holy.

There are many like this who do not wish to be confronted with a holy and righteous God who is the absolute standard of what is right. These people, in their desire to be able to do whatever they want regardless of how destructive it is, will not only deny that there is an absolute right but will also deny the God who is the absolute and who has planted his standard of right and wrong within us all.

Nevertheless it is important for us all to be able to give a logical, systematic and biblical defense of what we believe that we may be always ready to give a defense of the hope that lies within us (I Peter 3:15).

Read more about it

Anderson, Dr. Jim, *Christian Apologetics A Defense Of The Faith*

Evans, C Stephen, *Philosophy of Religion*

Hall Verna M. ed. *The Christian History of the Constitution of the United States of America*

Kennedy, D James, *Why I Believe*

Miller, Darrow L, *Discipling Nations*

Morris, Thomas V. *Our Idea of God*

CHAPTER EIGHT

FROM LOCAL TO GLOBAL

When I was a young man I was discipled by a man
who also discipled men who were, in today's lingo,
international students. One of them became a chief
member of the government of Egypt and another was of
the royal house of Japan. By leading these men to the Lord
and by teaching them the Bible and basic doctrine he was,
at the same time, impacting the world.

Leading "International students" to the Lord is one
obvious way to impact our world by what we do locally.
However, there is a sense in that all we do locally will
impact the world globally. There is a principle which
simply states that our theology dictates our philosophy
and our philosophy will dictate our society. This means
that our theology will impact our sociology as what we
think will become what we do. This has implications
for disciplines such as science, economics and education.
When the United States began with proper theology
we became a great nation that was the world's leader in

science, economics and education. As we now continue down a path away from proper theology we are, in equal proportion, losing our place of world influence in these same three disciplines. To put it in the words of Darrow Miller:

> The universe is intelligible. "The God who is there" is rational. He created an orderly universe and made man in His image. Because of these facts, truth, wisdom, and freedom not only are possible but also are to be treasured. These spiritual values have a direct impact on the material world in science, economics, education, and the "civilizing" of society. Those who wish to help the poor and disciple the nations must be "wisdom workers," helping the poor to rewrite their stories—and their lives."[30]

The fact that Christianity (proper theology) had a profound and positive impact on the founding and the future prosperity of the United States is well documented. For example a French social philosopher by the name of Alexis de Tocqueville visited America in the 1830's and made the following observations:

> The sects that exist in the United States are innumerable, ... They all differ in respect to the worship which is due to the Creator; but they all agree in respect to the duties which are due

---

[30]   Darrow L. Miller, *Discipling Nations* (Seattle: YWAM publishing, 1998) 119.

from man to man. Each sect adores the Deity in its own peculiar manner, but all sects preach the same moral law in the name of God . . . . Religion in America takes no direct part in the government of society, but it must be regarded as the first of their political institutions; for if it does not impart a taste for freedom, it facilitates the use of it . . . . How is it possible that society should escape destruction if the moral tie is not strengthened in proportion as the political tie is relaxed? And what can be done with a people who are their own masters if they are not submissive to the Deity?[31]

For the Americans, the ideas of Christianity and liberty are so completely mingled that it is almost impossible to get them to conceive of one without the other; it is not a question with them of sterile beliefs bequeathed by the past and vegetating rather than living in the depths of the soul.[32]

It is worthy of note that our founding fathers, including those who signed the Constitution were either Christian or at least Deist who had confidence in the Scriptures as the Word of God and believed that the teachings of Scripture are necessary for our type of democracy to

[31] Russell Kirk, *the Roots of American Order* (Wilmington Delaware: ISI Books, 2003 4th ed.) 332-333.
[32] Ibid. 448.

survive. "Virtually all of the 55 writers and signers of the Unites States Constitution of 1787, were members of Christian denominations: 29 were Anglicans, 16-18 were Calvinists, 2 were Methodists, 2 were Lutherans, 2 were Roman Catholic, 1 lapsed Quaker and sometimes Anglican, and 1 open Deist—Dr. Franklin who attended every kind of Christian worship, called for public prayer, and contributed to all denominations."[33]

Note the words of our first Supreme Court Justice John Jay:

> I recommend a general and public return of praise and thanksgiving to Him from whose goodness these blessings descend. The most effectual means of securing the continuance of our civil and religious liberties is always to remember with reverence and gratitude the source from which they flow.

> The Bible is the best of all books, for it is the word of God and teaches us the way to be happy in this world and in the next. Continue therefore to read it and to regulate your life by its precepts.

> [T]he evidence of the truth of Christianity requires only to be carefully examined to produce conviction in candid minds . . . they who undertake that task will derive advantages.

---

[33] William Federer, *America's God and Country Encyclopedia of Quotations* (St. Louis: Amerisearch, INC., 2000) 153-154.

Providence has given to our people the choice
of their rulers, and it is the duty as well as the
privilege and interest of our Christian nation, to
select and prefer Christians for their rulers.[34]

Another statement of interest comes from a man who
was extremely influential in the founding of this country,
John Adams: SIGNER OF THE DECLARATION
OF INDEPENDENCE; JUDGE; DIPLOMAT; ONE
OF TWO SIGNERS OF THE BILL OF RIGHTS;
SECOND PRESIDENT OF THE UNITED STATES

The general principles on which the fathers
achieved independence were the general principles
of Christianity. I will avow that I then believed,
and now believe, that those general principles of
Christianity are as eternal and immutable as the
existence and attributes of God.[35]

The list of quotations from great men can go on
forever, in fact there is a book written which contains 710
pages of quotes which is worth the purchase for those who
wish to read more about it, *America's God and Country
Encyclopedia of Quotations* by William J. Federer. A web
site that is valuable in this regard is www.wallbuilders.
com.

---

[34] David Barton, *The Founding Fathers on Jesus, Christianity and the
Bible* [article online] (accessed 6 October 2010) available from
http://www.wallbuilders.com/LIBissuesArticles.asp?id=8755.
[35] Ibid.

So what's the point and how does it relate to Evangelism? It is clear to any average thinking person that our country has moved away from these Christian/biblical principles which made it a great and prosperous nation. In our prosperity we have not only enriched the rest of the world but have taught the countries of world how to be prosperous and self-sufficient in their own right. An overwhelming amount of the good that America has done on behalf of the world has been accomplished through the agency of our missionaries. **It must be understood that this prosperity is in direct correlation with our theology. Therefore, America must not only remain prosperous, America must remain Christian**. The influence of the Church in the world is in direct proportion to the influence of the Church in America. This then creates an intimate connection between local and global evangelism. While we consider our vision for reaching the world we must also consider what we must do to turn our own country back to the Word of God. In previous chapters we have seen how the gospel changes people from the inside out and how these changed people will change their culture and then their society. Therefore, our first step in global evangelism is to give the good news of our Lord Jesus Christ at every possible opportunity i.e. evangelism. Now let us give some consideration to how we might think of global evangelism as an extension of our local evangelism.

I once attended a conference sponsored by a well respected mission agency. The conference addressed some interesting themes that have actually been around for at least the last ten years but are now becoming more pronounced. It was stated in the conference that there is a movement of the Spirit of God among Churches to step up and take a more proactive role in missions and the end result may be that Mission Agencies who do not get on board with this movement will cease to exist. This movement is based upon the premise that due to the fast changing world in which we live we can no longer do missions in the same way that we have been doing it for the last 100 years. Therefore, the local Church must make radical changes to their missions program to get up to speed with what God is doing in our world of today.

The traditional view of missions can be summarized with the phrase "**Pray, Pay and stay out of the way**". For at least the last 100 years, this has been the role of the Church in missions. The Church is to pray for missionaries which means having their pictures and bios on a wall some where and having occasional reminders for folks to pray, including an annual mission's conference. The second thing is to pay or provide financial support for missionaries. This is usually done in a random manner without any specific goal or focus on whom or what people group the Church is trying to reach. This is known as the shotgun approach where money is scattered out all over the world with no specific target or interest. Then the Church is to stay out of the way. Recruitment is done by the Mission Agency, training for missions is done by the Bible College, Seminary and Mission Agency, accountability, emotional/spiritual support and administration is all done by the Mission Agency. The Church is permitted to write letters send emails and an occasional gift box, but other than that it is best for the Church to stay out of the way and let the professionals take care of things.

However, the Lord is doing something new in Churches and certain agencies today. The new model is summarized as "**Pray, Pay and Stay**". Churches all across this nation are insisting that they have a key role in who goes where and what people group the Church is going to reach. This means that the praying and paying is now focused on a particular people group and the team that has been sent from the Church to reach them with the gospel. The Church plays a key role in the development, training, accountability, and emotional/spiritual support of the team as the team will be people who have been raised up out of that local Church. The Mission Agency is used as a tool for such things as administration and specific training, to be used by the local Church to accomplish the set goals of the Church.

One of the first things that we learn about global evangelism in the book of Acts is that local becomes global. There is an intrinsic link between what happens in the local Church and what happens in the world. This is in keeping with the words of our Lord that we are to be His witnesses, first where we are and then in an ever widening circle of influence until it reaches to the end of the earth, Acts 1:8. It is most unfortunate that this key component to global evangelism has been lost in the modern day method of doing missions. However, as I mentioned in the paragraph above, this is changing. This new movement goes back to the original pattern as seen in the book of Acts. In this pattern the local Church is an essential player in global evangelism and not just the source of funding.

The book of Acts is the first book written about the history of the early Church and is called the book of the Acts of the Apostles, the emphasis is on what they did more than on what they said. Doctrine and dogma is important but that alone is not enough, what we know must be followed by what we do James 1:22. So after we study our doctrine of evangelism (soteriology) and method for evangelism, we need to be doing evangelism. While our doctrine never changes, our methods are always changing according to the times and circumstances. The manner in which we did global evangelism 60 years ago is no longer relevant in today's fast changing world.

It is important to note that the word "missions" is not found in the divinely inspired Scriptures. The word comes from the Medieval Latin word *mittere* which means to send. The word is more commonly used of the place to which a person is sent i.e. a mission or the person being sent i.e. a missionary. It makes sense that the word would come from the Latin as the first missionaries were Roman Catholic Monks (Roman Catholic and Nestorian around A.D. 500). Their job was to build a mission and convert the heathen to Christianity (Rome's version). In contrast to the words missions, missionaries, and mission boards, the Bible uses words such as gospel (ευαγγελιον), evangelist (ευαγγελιστης) and ministry (διακονος). At the same time it is true that the use of the words mission, missionary etc. have been adopted by the Church of today and the common use of these words have made them acceptable for dialogue in today's common English. However, it is important to note that there is nothing sacred or even inspired about either the terminology or the method.

Now back to the Scriptures to be reminded of what took place. In **Acts 13**, it was as Paul and Barnabas ministered to the Lord that the Holy Spirit said, "Now separate to Me Barnabas and Saul for the work to which I have called them." (Acts 13:2). This "call" **was a result of doing the work** which they would continue to do, only in Asia Minor. It was the Church at Antioch who prayed and fasted and then determined that it was of the Holy Spirit that they let them go. From then on the book of Acts can be outlined by the cities that Paul visited and preached the gospel. He did not go to one place and stay there for life or until retirement but went from one city to another. It was expected that each Church that was started would then raise up elders and that the Saints of each Church would recognize and use their spiritual gifts to establish the Church and plant even more Churches in the area. For example we understand that the Church in Ephesus sent out those who then planted churches in Colosse and Laodicea and others as well. The making disciples of the nations were the direct result of the overflow of the working of the Holy Spirit in the life of the believers to preach the gospel from one community to the next etc. The result is that the whole world was heavily influenced by the Church.

In **Acts 14:26** we read how Paul returned to the Antioch Church "where they had been commended to the grace of God for the work which they had completed." Notice how this stands in contrast to the "long-term" or "career" missionary. **The Church had commended him**

to the grace of God for the work, not to be a "missionary society" nor is Paul the "general director" or "coordinator". In fact, Paul is an Apostle with the authority of an Apostle but conducting himself as a servant of Christ. The terms Apostle and servant are the actual words to describe Paul in the Scriptures. He then stayed with the disciples many days until he and Barnabas were asked to go to Jerusalem to argue the question concerning circumcision. Once again, Paul is not seen as working independently of the Church but **rather as one working in cooperation with the Church and at their request.**

In **chapter 15** Paul was successful in his argument and returned to Antioch and the Church. Paul and Barnabas remained in Antioch teaching and preaching the word. They then determined to go back and visit the brethren in the cities where they had been to see how they were doing. Paul then disagreed with Barnabas as to who should go and so Barnabas took Mark and sailed to Cyprus while Paul took Silas and went to Syria and Cilicia to strengthen the Churches. It is difficult to see Paul as the "general director" when his right hand man does not agree with him and goes in a different direction. Once again they were commended by the Church to the grace of God. In stark contrast to being separate entities, the evangelists work in full cooperation with the Church and it is the Church who commends them to the grace of God.

In **chapter 16** of Acts Paul had Timothy go with him, but first had him circumcised. This was in spite of the fact that Paul had won the argument that there was no need for Gentiles to be circumcised, but because his mother was a Jew and to keep peace with the Jews he had Timothy circumcised. This is hardly the act of one who is acting independently of any other authority. The book of Acts continues with the travels of Paul and of his team. The key word is "travel" he does not stay in one place and many times had to run for his life. But he planted Churches and then moved on to plant more Churches. The idea of setting up a headquarters from where he might oversee his "mission's empire" is totally foreign to the text.

It would now behoove us to examine the Epistles to see if we can discover any direct doctrine which would give a biblical defense of the traditional mission's model. This is important because it is from the Epistles where we get our doctrine i.e. line of teaching for how we do Church. The Gospels give us the gospel of our Lord Jesus Christ of which the greater part is committed to His death, burial and resurrection. The book of Acts gives us the acts of the Apostles during the first 100 years of the Church. The Epistles teach us how we should then live in this present evil world and how we are to do ministry where a large portion of that is the preaching of the gospel. But when we examine the Epistles we do not find any instructions that in any way resemble the traditional mission's model. So what do the Epistles teach about making disciples of the nations?

It is interesting that in the modern/postmodern Church there is generally an emphases placed on **giving to missions**. Yet in the Epistles the most lengthy dissertations on giving had to do with giving to the church at Jerusalem which had become impoverished partly do to their experiment in communalism, which has never worked, and due to the famine in the area. The only other mention of giving would be to those who are preaching the gospel as one who preaches the gospel should be able to live from the gospel (I Corinthians 9:7-18) and the elder who is worthy of double honor i.e. a wage (I Timothy 5:17-18). **Notice how instead of the home Church sending money to the Church plants, the Church plants were sending money to the home Church. In fact, very little if anything, is said about the support or sending money to missions in the New Testament**. The New Testament Church seemed to follow the instructions of our Lord in Matthew 10:9-10, "Provide neither gold nor silver nor copper in your money belts, nor bag for your journey, nor two tunics, nor sandals, nor staff; for a worker is worthy of his food." This was done to the extent that the Apostle Paul would ply his trade to

support himself and apparently those who were with him
(I Corinthians 4:12; 9:15-18).

So what do the Epistles **teach concerning the
spreading of the gospel and making disciples of the
nations?** Certainly the Epistles did not remain silent
concerning the command of our Lord and there is a great
deal of evidence that the gospel did, in fact, travel to the
far corners of the world. What we find in the Epistles are
instructions as to how and why we should do ministry.
Let us examine a couple of examples of what is meant.

It begins with our salvation which is by grace through
faith in our Lord Jesus Christ. Immediately following is
good works as in Ephesians 2:8-10 and Titus 3:4-8. We
have been placed into a sphere of good works which God
has before ordained that we should order our conduct
accordingly. We then are lights in this world of darkness
Matthew 5:16. A key and necessary component of these
good works is the ministry of reconciliation I Corinthians
5:17-19. This ministry of reconciliation is not confined
only to the professional full time worker but is to be
carried out by all who have been reconciled to God. The
reason why every believer is given a gift or gifts from the
Holy Spirit is to do the "work of the ministry" Ephesians
4:12. The Church of the New Testament understood and
practiced this doctrine. The result was that the gospel
spread from one community to another, to another, to

another, to the uttermost parts of the world. It was as if the grace of God over-flowed each believer out to others who could then also experience this grace and in turn over-flow to still others.

Change in any Church is very difficult, the status quo is maintained just because it is the status quo i.e. "we have always done it this way". So it begs the question, even if we were inclined to move toward a new model of local to global evangelism, how would we do it? The following is submitted as a starting point.

1. **We need to place greater value on the local Church**.

This means that we need to place greater value on the role of the local Church in evangelism. The old excuse that we have para-church organizations because the Church is not doing their job is no longer acceptable. Rather the Church must step up to its responsibilities and do the job.

A central theological reality is that the church is uniquely equipped to be the locus of missions because it is essentially missionary by its very nature. This means that the church itself is the missionary reality that God sends into the world—*it is far more than an institutional source from which funds and missionaries are sent or agency-developed programs carried out.*36

---

36 Ibid., p. 74

Obviously para-church organizations will always exist including missions agencies. The point here is that **the Church** (by Church the local Church in particular is in view) needs to take more **initiative and be more involved in the vision, direction and over all ministry of the agency**. This is particularly true of the missionaries who are being supported by the local Church. Presently the mission's agency move personnel from one place to another and from one position to another without any input from the Church. The agency simply sends out a form letter informing the Church of their decision and the Church is expected to like it and support it. This letter which informs of a move in position is usually accompanied by a plea for money as the Church has been relegated to a place of passive support on demand.

One of the problems that have contributed to this dilemma is that most missionaries must be supported by many churches, 7.6 on average [37]. Each of these churches then has a list of missionaries that they "support". It is like a badge of honor to have a long list of supported missionaries who are proudly displayed on a bulletin board of some kind. The fact is they do not actually support anybody, they are merely part of a collection of individuals and Churches who contribute to

---

[37]   Mark A. Noll, *The New Shape of World Christianity*, (Downers Grove, IL. IVP Academic, 2009) 118

their support. Therefore, if the agency was to get input from a local Church they would have to solicit a long list of Churches and hope that each one responds and each one has the spiritual discernment etc. to give credible input. This is another problem with the traditional mission's model that needs to be fixed with a new model.

In this new model, the local Church will focus its attention and support on one or a few evangelists that they are able to fully support. For example, a Church budget has about $100,000 committed to corporate support. Assuming that the average cost of supporting one evangelist and their family is $50,000 per year [38], this Church **could fully support two evangelists**. Other small Churches that do not have this kind of budget could band together with one Church acting as the spokesman for all of them on behalf of the evangelists they support. The ideal scenario would be of a growing Church which reaches its overflow level and then plants other Churches, these Churches together would then be able to fully support several evangelists. For example **if a Church were to triple in size then it could triple the number of evangelists it supports**. Church growth locally will then impact both local and global evangelism. Then as more Churches are planted they will be able to contribute

---

[38]   Haggai Institute, http://www.haggai-institute.com/

to the support of evangelists as well, while working in cooperation with the mother Church.

In this model the Church will have to be more particular about who they send out to the field and what it is that they are to accomplish. The criteria would consist of **someone who is trained by the local Church (with or without the help of a Bible College or Seminary) and who has proven themselves in the local Church as having the gifts and abilities to perform the task required in their particular ministry.** This would be consistent with the pattern of Acts 13 where Paul and Barnabas were involved in that ministry and then were let go by the Church to Asia Minor under the direction of the Holy Spirit. Another possibility for support would be that of **the native pastor and the Church already present in the country** and the support of those schools and training centers who train those native pastors. It is understood that these training centers would be well researched as to curriculum, doctrine etc. before support would be extended.

The agency would then be of assistance to the Church in getting the evangelist into the country, providing language study if that was not otherwise available, and other administrative task. However, the purpose, vision and goals of the evangelist would be consistent with the vision and goals of the local Church as they have been led by the Holy Spirit. The Church will provide the support, direction and guidance for the ministry to be accomplished, how long it should take and any changes that will be made in the ministry. Both the evangelist and the agency will be answerable to the Church. This will enable the Church to focus on that small group of evangelists to support, encourage, pray for and guide through their ministry. The ultimate goal is for the evangelist to plant a Church that will grow in the same manner as the mother Church which will in turn plant other Churches and thereby spread the gospel throughout that country, making disciples of the nations.

An important element to this model is that the evangelist who travels cross culturally must not combine Western culture with the gospel. **It is important for the evangelist to assimilate the culture of those he is trying to reach as much as possible**. It is true that this may create a life threatening life style but these are the potential sacrifices of cross-cultural ministry which must be carefully considered.

It needs to be noted that the term "missionary" has been replaced with the word "evangelist".

The reason for this is twofold, first it is consistent with the language used in the inspired text as noted above. Second, the term missionary has degenerated into anyone who goes overseas, whether they actually preach the gospel or not. The person can be a support staff of some kind and sits in an office, school, clinic etc. Generally speaking most of these "support staff" personal could be replaced by a national worker who is able to be supported at much less cost than the "missionary" and could really use the job to support his family. At the same time it is understood that we are not always talking about a "full time" evangelist. Remember that the Apostle Paul made tents to help support himself and his companions. But this does mean that whoever goes for the purpose of planting a Church needs to be someone who has proven their ability to give the Gospel and to disciple a new believer in the context of their home Church.

2. **We must adjust our attitude toward the peoples of the world.**

The days of the great white missionary being the only one qualified to bring the truth to the heathen have long since passed. There are many people who have come to the opinion that the descendants of Japheth are not the only ones capable of making disciples of the nations. In the first century it was the descendants of Shem and now it is the descendants of Ham who are becoming the majority of Christians. "The center of Christianity is no longer the northern hemisphere; it is the southern hemisphere" [39].

---

[39] Jim Eckman Issues in Perspective, www.issuesinperspective. com August 8-9, 2009;

"Of the approximately **2 billion Christians in the world today,** 648 million (11% of the world's population) are Evangelicals or Bible believing Christians, Evangelicals have grown from only 3 million in AD 1500, to 648 million worldwide, **with 54% being non-whites.**"[40].

Therefore, it is important that the evangelist of today be aware of the Church already in that culture or people group and be prepared to work alongside of that existing ministry. We dare not march into a people group and assume that we are the answer to all the problems and the only source of truth and salvation. But rather come with the servant mindset of how can I be of service to you? **The American Church must position itself into a support role for ministries already growing in the cross-cultural environment**.

In a place where it is believed that **no evangelistic witness exist**, there is still no need for the American Church to think that we are the only ones qualified to bring the gospel to that unreached people. More often than not there is a Church in a neighboring country or region with a similar people group with a similar culture and language. I once knew a person from Africa who could speak 8 languages which meant that she could travel to several countries around her home and be able to communicate.

---

40 Christianity Today - General Statistics and Facts of Christianity **http://christianity.about.com/od/denominations/p/ christiantoday.htm**

This is quite common in other countries. The result is that a nearby Church could more easily plant a church in the un-churched region because they may already know the language, are familiar with the customs and can live in that climate and environment without undo difficulties. Whereas an American will need to spend a great deal of time learning the language, customs and culture and may never be able to adjust to the climate and environment. In such a case it would be better for the American Church to assume a supporting role of the church in the foreign country who is better equipped to reach the neighboring region with the gospel. Once again this new model is more in keeping with the Church of the New Testament where one Church would reach out and plant another Church which would reach out and plant another and so on and so on until the gospel advanced from one people group to another, making disciples of the nations.

Not only is this new model more effective in today's world and more biblical but it also more cost effective. To fully support a national worker is about 1/10 the cost of supporting an American in a foreign country. **The Haggai Institute points out that, "Supporting organizations who train and support nationals for the work of the ministry is more cost effective, culturally sensitive, language/communication effective, provides access to closed countries, and provides greater safety and** security."[41]. America is no longer the world's great piggy bank, available funds to support the traditional missionary is fast diminishing which is seen in the loss of support being experienced by many supported missionaries.

3. **We must work with the understanding that evangelism begins at home and proceeds out from here**.

The great commission in Matthew 28:18-20 is not just for a select group of missionaries, it is for everyone who is a disciple of the Lord Jesus Christ. In Acts 1:8 our Lord gave further instructions as to the great commission, "but you shall receive power when the Holy Spirit has come upon you; and you shall be witnesses to Me in Jerusalem, and in all Judea and Samaria, and to the end of the earth." In following the reading of Acts we see that is exactly what happened. The gospel moved out from Jerusalem and continued to expand and grow until it had reached out into all the known world. It is interesting that there are **characters in the Mandarin alphabet that can trace their origins back to Christianity.** In our day we have badly underestimated the impact and influence of the first century Church. This does not mean to say that there are presently no unreached people in the world. It simply means that during their time the first century Christians took the gospel to the far reaches of the world

---

41   Haggai Institute, http://www.haggai-institute.com/

that they knew by planting one Church at a time and then allowing that Church to grow and then plant another. The idea of leap frog evangelism is of modern origin and not biblical. **One pastor has told his congregation that to be involved in missions is to plant a Church**.

We must focus on **being a Church of irresistible influence** where we have the understanding that what we do locally will impact the world. To be a church of irresistible influence means to be a light shining in the darkness. It means doing the good works that God has before ordained that we should order our conduct in them **Ephesians 2:10**. When others see our good works they will glorify God **Matthew 5:16** and will listen to the good news we have for them. Then the irresistible grace of God will draw them into this great salvation. They in turn will impact the lives of those in their nest of influence and gradually these will be added to our Church which will grow to the place of over-flow. At this point our Church will continue this process along with the new Church that has been planted. **In the mean time the process is going on in countries and in people groups all over the world until these Churches start connecting.** We will be called upon to provide resources and expertise to these Churches world wide with a view toward seeing the whole world blanketed with Churches. In this process we will be training one another and future generations to do ministry. **This is the key to our training and education that we teach the Bible and train how to do ministry so that no matter where a person finds themselves in the world they will know how to do ministry and teach the Bible**. The essential component to our ministry is to know how to give the gospel and be so comfortable giving the gospel we will do it at every opportunity, anywhere in the world and in any language.

An impossible task?

At first glance this appears to be an impossible task because there is a great deal to overcome to get to the place of the new model. There will need to be a change of mindset where the biblical merits of the new model can be understood and embraced, that this is in fact a worthwhile goal. To move from where we are to where we could be will require incremental steps beginning with setting more value and importance on the local Church, then thinking differently about the peoples of the world, and finally begin making disciples where we are with a view toward expanding out to the rest of the world.

Read More About It

Engel, James F. & Dyrness, William A. *Changing The Mind Of Missions, Where Have We Gone Wrong?*

Federer, William, *America's God and Country Encyclopedia of Quotations*

Kirk, Russell, *The Roots of American Order*

Miller, Darrow L., *Discipling Nations*

Noll, Mark A., *The New Shape of World Christianity*

Nyquist, J. Paul, *There Is No Time*

CHAPTER NINE

THE LOCAL CHURCH

While growing up in a Christian home it was the practice of our family to attend Church on Sunday and Wednesday night. We were a part of the Plymouth Brethren and did not have a paid clergy, therefore the teaching was done by men who ostensibly were gifted by the Holy Spirit to teach and the pastoral care was done by elders (at least theoretically). Within this framework we had a Worship Meeting during the first hour on Sunday morning (to bring our firstfruits to the Lord) and this was followed by Sunday school and the Adult Bible Hour. We celebrated the Lord's Supper every Sunday during the Worship Meeting. By the time we left the chapel we had spent about three hours in Worship, Bible study, Fellowship and Breaking of Bread. Sunday night was the time for a gospel meeting for another hour. It was called a gospel meeting in spite of the fact that only saints and seats were in attendance. On Wednesday night we spent another one and half to two hours in prayer and

Bible Study. When I became a teenager, we formed our own version of a Youth Group before Youth Groups had become popular. We met on Sunday afternoon which then added another two hours to our time in Church. During that time we sang together, had a brown bag supper and a Bible Study. It was in this framework that I received my Bible education, this along with older men who took an interest in me and "mentored" or "discipled" me with more Bible study and practical experience. I served the Lord and His people for over twenty years before learning that I needed to be credentialed and then went to a Bible College to earn my Bachelors Degree in Philosophy, History and Bible which led ultimately to a Doctor of Philosophy in biblical studies.

The point here is that Church was very important to us. When I became a teenager my Father decided it was time for me to make my own decisions and told me that I did not have to go to Church (*meeting* in the Brethren vernacular). At the time I thought that was the oddest thing that my Father could possibly say, I could not even conceive of not going to Church and not being heavily involved. It made as much sense as if he had told me that it was alright for me to smoke cigarettes and drink whisky. Going to Church was an absolute essential part of my life; it was more than mere routine, it was as important as breathing, eating and drinking. I would no more skip Church than skip dinner. When I went to play baseball as soon as I found out that they played on Sunday, I immediately quit the team. I never took a job that required that I work on Sunday or Wednesday night, those times were reserved for Church. Unfortunately we now live in a day and age where the Church has lost its importance and its value in the lives of the Lord's people. In this lesson we will do a brief study of the Church and why it is of value particularly in evangelism.

Any study of the Church will need to begin at the first mention of the word Church in the Bible which would be Matthew 16:18 ". . . and on this rock I will build My church, and the gates of Hades shall not prevail against it." With only one exception the word *Church* is translated from the word εκκλησια (ekklasia or ecclesia) which literally means called out ones or assembly. When we study ecclesiology we are doing a study of the Church.

There are at least two important points to note in the above stated text. The first is that our Lord has clearly stated that He will build the Church. We often make the mistake in thinking that we are to build it, our expertise in marketing, methodology and programming is what builds the Church. We tend to look for the latest fad or new thing that will draw in the crowds of people. The simple fact of the matter is that the Lord will build His Church and He will do it the same way that He has been doing since the first century i.e. His people will preach the gospel, the Holy Spirit will save them and they will come to Church. Studies will show that people are more apt to come to Church because their lives have been transformed through the Gospel and because someone who cares invited them.

> For the church, winning is the transformation of people and communities in every part of God's world! (Matt. 28:19,20; Acts 1:8)[42]

> If your church is transforming people, reproducing at every level and staying true to the unique vision that God has given it, we all win, attendance will take care of itself.[43]

> . . . Pastor Lee Powell of CedarCreek made the comment that despite all the advertising they do (he compared it to buying weekly tire ads for Sears), **80% of first time visitors come because they were personally invited.**[44]

> South Carolina, e-mailed us recently to say that his church relies primarily on people inviting other people—to the extent that they've almost

---

42 Dave Ferguson, "Winning at Any Size," *Outreach* 2010 Special Issue (September 2010) 134.

43 Ibid.

44 Kevin Hendricks, *Your Invited: Bringing People to Church* [article online] (accessed 13 October 2010) available from http://www.churchmarketingsucks.com/2005/04/youre-invited-bringing-people-to-church/.

given up on mass mailings.[45] Great suggestions! I work for <u>New Spring Community Church</u>, also in South Carolina. We have grown to over 3500 since starting in Jan. 2000. We spend money on billboards, newspaper advertising, mailers, etc . . . . and consistently the "persona <u>Shawn Wood</u>, the Creative Communications Pastor . . . at <u>Seacoast</u>, a multi-site church in 1 invitation" is the one thing people raise their hands to in our membership class as the reason they started attending.[46]

Take note of how personal evangelism is key to growth and when the Lord's people see the value of their local Church in transforming lives, they invite others to join them. The preaching of the gospel as a method for Church growth is seen throughout the book of Acts where the gospel is preached and the Church grows and multiplies e.g. Acts 2:41 three thousand were added; 2:47 and the Lord added to the Church daily those who were being saved; 4:4 about five thousand; 5:14 "And believers were increasingly added to the Lord, multitudes of both men and women," 6:7 "the number of the disciples multiplied greatly in Jerusalem," and on it goes. **There is nothing sacred or spiritual about being a small Church**!

The second thing that we learn about the Church from Matthew 16:18 is the place our Lord occupies in

---

45   Ibid.
46   Ibid.

the Church. He is the rock upon which He will build His Church. Simon is called Peter (Petros) a little rock, but it is on the big rock (Petra) that the Lord will build His Church. How do we know this? Our Lord referred to Himself as the stone rejected by the builders but that which became the chief cornerstone (Matthew 21:42-44). Peter understood what was meant in this conversation as that is what he preached in Acts 4:11-12 "This is the stone which was rejected by you builders, which has become the chief cornerstone. Nor is there salvation in any other, for there is no other name under heaven given among men by which we must be saved." Then again as he wrote about it in I Peter 2:4-8

> Coming to Him as to a living stone, rejected indeed by men, but chosen by God and precious, you also, as living stones, are being built up a spiritual house, a holy priesthood, to offer up spiritual sacrifices acceptable to God through Jesus Christ. Therefore it is also contained in the Scripture, "Behold, I lay in Zion A chief cornerstone, elect, precious, And he who believes on Him will by no means be put to shame." Therefore, to you who believe, He is precious; but to those who are disobedient, "The stone which the builders rejected has become the chief cornerstone," and "A stone of stumbling and a rock of offense." They stumble, being disobedient to the word, to which they also were appointed.

He is not only the foundation of the Church but He is also the head of the Church Ephesians 1:22; 5:23; He is the head of the body, the Church Colossians 1:18; and it is the head from which we receive our direction, nourishment and unto which we grow to maturity Ephesians 4:13-16. Therefore as we value Jesus Christ as Lord of our lives then we value the Church because of His position in the Church and His intimate relationship to the Church. This is the reason why we are taught by the inspired Scriptures to endeavor to keep the unity of the Spirit in the bond of peace. This is the reason why we have been equipped with gifts and abilities to minister in Church which is His body (see I Corinthians 12:12-31).

Next we see the value of the local Church in the pattern set for us in the book of Acts. The first mention of the Church in Acts is 2:47, ". . . praising God and having favor with all the people. And the Lord added to the church daily those who were being saved." This text begins with those who received the word and were baptized. There were three thousand souls added to the group of one hundred and twenty disciples. Notice how we have three thousand one hundred and twenty in the Church of Jerusalem and the Lord adds to that number "daily those who were being saved." Today this would be called a mega church and growing. This Church was known for four things; doctrine, fellowship, breaking of bread and prayer (verse 42). Then they were also known for having all things in common so that they met the needs of all the Church. So what we have so far in the Church is that there is a large number who respond to the Gospel on a daily basis, they spend time together in doctrine, fellowship, breaking of bread, prayers and they cared for one another both spiritually and physically. This wonderful community of believers begins and continues to grow in **direct correlation to evangelism**.

As we continue through the book of Acts both evangelism and Church growth continue. In chapter 4 we see that those who heard the word believed and another 5,000 were added. After Peter is arrested the Church gathered to pray and the place where they were was shaken. The Church continued to care for one another as they had all things in common and there was not anyone among them who lacked. In chapter 5 those who lied to the Holy Spirit fell dead and were carried out and fear came upon the Church. In chapter 6 there was a problem with the distribution of goods, this problem was solved with the appointment of Deacons who were men full of faith and the Holy Spirit. When Peter went to the Gentiles in chapter 10 he was held to account by the Church in chapter 11. Accountability was accomplished by the Church and no one, even Peter, was allowed to go off on his own without reporting back to the Church.

A key chapter in the book of Acts is chapter 13 where the Church at Antioch becomes the spring board for taking the gospel even further into the uttermost parts of the world. It is the Church at Antioch which is able to see the ministry of Paul and Barnabas so that when the Holy Spirit directs the Church to separate them for the ministry for which they have been called, then it is the Church who turns them loose for this work. It is important to note that Paul and Barnabas were not called by the Spirit outside of the Church and then went to the Church for support but rather while working in the Church, the Church let them loose to go to this new ministry which would be an extension of the ministry of the Church at Antioch. No mention is made of finances, support etc. they were simply sent out after prayer, fasting and the laying on of hands. Upon the completion of this journey, they returned to Church at Antioch to give their report. Paul was not on his own, nor did he form some separate entity apart from the Church but rather kept himself in submission to the local Church of Antioch.

In continued submission to the Church Paul and Barnabas are sent to Jerusalem to argue the question of circumcision in chapter 15 and they return to Antioch with the decision made at the council in Jerusalem. The remainder of the book of Acts is committed to the travels and ministry of Paul. He plants Churches, organizes them with elders and deacons and moves on. Paul only spends a few months in each of the Churches that he plants. The longest that he stays in any one place is three years in Ephesus and even then he was moving about seeking to encourage and strengthen the other Churches in the area. He spent one and half years in Corinth but other than that we can see how he spent but a few months in any one Church. This would raise the question as to how could he do that? How is it that these Churches were able to continue on without the Apostle Paul there to guide them, to mentor them, to disciple them? It would seem that Paul had a great deal of confidence in the Lord building His Church, in the power of the Holy Spirit to teach and hold them up and upon the Scriptures which in time included his own Spirit inspired letters. There might be a lesson in this for us in our modern times. At the end of Paul's first two journeys he reported back to the Church at Antioch from where he had been commended to the grace of God and His work. At the end of His third journey he went back to Jerusalem. It is to the Church that Paul was accountable and he never forsook the authority of the Church in his life and ministry.

Next we want to consider the emphasis placed on the "Local Church" in the Scriptures.

By doing a simple count of the word "church" in a Strong's concordance we find that it is translated from the Greek word *ekklesia* 116 times in the New Testament, of those 36 are in the plural. A count of the Greek word *ekklesia* in a Greek Concordance shows the word being used 115 times but 3 of those times it is translated "assembly" in Acts 19. Generally it is argued that the Church is seen as both universal and as local, "We note, then, positively what the church is. The term "church" is used in two senses: the universal sense and the local sense"[47]

This leads us to a number of questions as to how we define the word, how it is used in Scriptures, and what is meant by its use? This word is not unique to the Scriptures, our Lord did not make it up, but instead used a word that was in common use. A Lexicon definition of the word *ekklesia* will include both the common usage of the word and then how it was used in Scriptures:

> Among the Greeks . . . an assembly of the people convened at the public place of council for the purpose of deliberating:
>
> . . . a company of Christians, or of those who, hoping for eternal salvation through Jesus Christ, observe their own religious rites, hold their own religious meetings, and manage their own affairs

---

47   Henry C. Thiessen, *Lectures in Systematic Theology* (Grand Rapids: Eerdmans Publishing, 1979) 310.

according to regulations prescribed for the body
for order's sake;[48]

Wolston in his lectures on the Church defined it this
way, "The Greek word ekklhsia here translated "church"
in our version, meant originally *an assembly of the citizens
of any particular state.*"[49] Another commentary defined
it as "an organized assembly, whose members have been
properly called out from private homes to attend to public
affairs. This definition necessarily implies prescribed
conditions of membership." [50]

We can summarize the meaning of the word by saying
that it is a specific group of people called out to a specific
place for a specific purpose at a specific time. This begs the
question, by definition, how can a Church be universal?
In certain circles at first gasp, this question can seem to be
heretical. However, the point here is that in the Scriptures
the way in which the word is used most often is, by far,
the local Church. The concept of the universal Church
is contrary to the definition of the word because it is not
possible for a "universal Church" to all meet together,

---

48   Joseph Henry Thayer, *Greek-English Lexicon* (Grand Rapids: Zondervan, 1974) 196

49   W.T.P. Wolston, *The Church: What is it?* (Denver: Wilson Foundation, 1982) 15.

50   Weldon Adams, *The Church According To The Scripture* (Shreveport: LinWel Publishing, 2000) 15.

with the exception being when we all meet together in glory.

To understand the concept of the "universal Church" we need to take note of those passages of Scriptures that seem to teach this concept. The first of these is the one already discussed in the paragraphs above Matthew 16:18 where the Lord will build His Church. This certainly includes every local Church in every location in the entire world. Therefore, when people anywhere in the world come to Christ by faith and begin to gather together, they are the Church and the Lord has and will continue to build that Church. In Ephesians 1:22-23 we read, "And He put all things under His feet, and gave Him to head over all things to the church, which is His body, the fullness of Him who fills all in all." Here the word "Church" is qualified by the phrase, "'which is His body' as the word 'which' is *hetis,* 'which is of such a nature as', and has a qualitative nature to it."[51] The "qualitative nature" here is that which is being spoken of as the Church is the body of Christ so that the "universal Church" is spoken of in terms as being the body of Christ. When a person is born again they become a member of the body of Christ and as they attend a local Church that Church is under the authority of Christ and His Word as He is the head of the

51   Kenneth S. Wuest, *Wuest's Word Studies* (Grand Rapids: Eerdmans Publishing 1973) 56.

Church, His body. Again in chapter five of Ephesians we read, ". . . as also Christ is the head of church; and He is the Savior of the body." In Colossians 1:18, the order is reversed but the meaning remains the same, "And He is the head of the body, the church," the universal sense is again represented by the concept of the body.

In I Corinthians chapter twelve we have a most thorough discussion on the body of Christ. This discussion begins with verse 12 where we see that there are many members in this one body, and that these many members become a part of this one body through the activity of the Holy Spirit as seen in verse 13, "For by one Spirit we were all baptized into one body". All the members of the one body have equal value in the body while having different functions verses 14-25. Since all the members are linked together in this one body they feel one another's pain and rejoice with one another's honor, verse 26. Now God has appointed different ones in the Church to be gifted in such a way as to benefit the Church verses 27-31. In verse 28 where we read, "And God has appointed these in the church . . ." this would certainly indicate the universal Church which would be true in the sense that this verse is connected to the whole teaching on the one body of Christ.

In the meat of the New Testament Epistles we see this qualifier of the universal Church being the body of Christ. The exception is the verse from the Gospel of Matthew 16:18 and then another exception is found in Hebrews 12:23 where we read, ". . . to the general assembly and church of the firstborn who are registered in heaven," Here the text is referring to the whole assembly of believers who are registered in heaven which would include all believers everywhere.

> Weldon Adams in his book *The Church According To The Scriptures* argues as follows:
>
> The Church of God is never used of any institution, except of an assembly or congregation of baptized believers in some given locality.
> The local church is the only kind of church God has on the earth. It is both visible and functional.[52]

So what do we do with the passages cited above which have a universal connotation? Adams explains as follows:

All the remaining verses using the word *church* (ekklesia) can be categorized into one of four usages:

1. Abstract—refers to the church institutionally.
2. Generic—refers to the church instructionally.
3. Prospective—refers to the church in glory or in prospect.
4. Local—refers to specific local churches.[53]

---

[52] Weldon Adams, *The Church According To The Scriptures* (Shreveport: LinWel Publishing 2000) 30.
[53] Ibid. 32.

The key word in Adam's argument is "functional" i.e. the abstract, generic, and prospective church is not functional. It is only the local Church which is able to evangelize or defend the faith. It is only the local Church which is able to assemble together for doctrine, prayer, breaking of bread and fellowship. It can be further argued that all the statements made which are viewed as the universal Church can be applied to the local Church. For example when we read that the Lord Himself will build His Church, this is true of every local Church everywhere and when we say that the Lord is the head of the Church, this is true of every local Church everywhere. Every local Church can say that the Lord is the head of this Church and He will build it.

This brings us to the great theological question of **so what**? As stated in the opening paragraphs of this chapter **the local Church has been devalued**. People no longer think of the Church as a springboard toward evangelism and ministry. Instead we have what is known as parachurch organizations which are the instruments for evangelism both local and global and for ministry in general. Since people view themselves as a part of the universal Church they see no need for the local Church to be their point of ministry. It is possible to only use the local Church for ones own purposes and then move to the parachurch for whatever type of ministry that suits them at the time. In order to proceed from this point, we will need an understanding of "parachurch" organizations and what they have done to the local Church.

The prefix *para* means beside, alongside of, beyond, or aside from[54] A "parachurch organization" therefore, is one that works either alongside of or beyond the local Church. The online encyclopedia defines a parachurch as in the following:

> **Parachurch organizations** are Christian faith-based organizations that work outside of and across <u>denominations</u> to engage in <u>social welfare</u> and <u>evangelism</u>, usually independent of church oversight. These bodies can be <u>businesses</u>, <u>non-profit</u> corporations, or private <u>associations</u>. Most parachurch organizations, at least those which are normally called *parachurch*, are <u>Protestant</u> and <u>Evangelical</u>. Some of these organizations cater to a defined spectrum among evangelical beliefs, but most are self-consciously <u>interdenominational</u> and many are <u>ecumenical</u>.[55]

You will note that a parachurch is one that engages in "social welfare and evangelism". The evangelism is not only local but global. There are hundreds of local and international parachurch organizations and more every day. Some of the most commonly known include Campus

---

54 *Merriam Webster's Deluxe dictionary,* 1994 10th ed., "para."

55 Parachurch Organizations, <u>http://en.wikipedia.org/wiki/ Parachurch organization</u> [online]

Crusade, Fellowship of Christian Athletes, Youth for Christ, Billy Graham Institute, Cadence International, Young Life, Navigators, Intervarsity, Crisis Centers, Good News Jail and Prison Ministries, Gideons and on and on it goes. In addition to these you have hundreds of Bible Colleges and Seminaries and then add to these hundreds of mission's agencies. So then how are all of these parachurch organizations supported? Theoretically, they are supported from discretionary income that is given over and above what is given to the local Church. The reality is that it is money taken from the local Church and given to causes deemed more worthwhile than the ministry of the local Church. The same can be said of the manpower to run these organizations i.e. it is people who rather than serve in the local Church will serve in the parachurch organization leaving the local Church, many times, under served.

An article found in a blog on the internet summarized parachurch in the following manner:

> Parachurch organizations are involved in mass evangelism, campus evangelism, urban mission, Christian education (K-12, college, seminary), political activism, legal aid, Christian media (print, radio, TV, film, music), crisis counseling, apologetic blogs, jail and prison outreach, drug rehab, Bible societies, home-schooling, home Bible study, &c.

Although they often cooperate with various churches, they are self-governing. This autonomy has made them controversial in some Evangelical circles where they are viewed as encroaching upon or usurping the Scriptural prerogatives of the church. They drain away resources. They're rogue elephants. They dilute doctrinal purity.[56]

William MacDonald, an often published writer, had this to say about parachurch organizations:

In recent years there has been an organizational explosion in Christendom of such proportions as to make one dizzy. Every time a believer gets a new idea for advancing the cause of Christ, he forms a new mission board, corporation, or institution! One result is that capable teachers and preachers have been called away from their primary ministries in order to become administrators. If all mission board administrators were serving on the mission field, it would greatly reduce the need for personnel there. Another result of the proliferation of organizations is that vast sums of money are needed for overhead, and thus diverted from direct gospel outreach. The greater part of every dollar given to many Christian organizations is devoted to the expense of maintaining the organization rather than to the primary purpose for which it was founded. Organizations often hinder the fulfillment of the Great Commission.

---

[56]   *Church or Parachurch* [article on line] (accessed 17 December 2010) available from http://triablogue.blogspot.com/2006/01/church-or-parachurch.html; Internet.

Jesus told His disciples to teach all the things He had commanded. Many who work for Christian organizations find they are not permitted to teach all the truth of God. They must not teach certain controversial matters for fear they will alienate the constituency to whom they look for financial support. The multiplication of Christian institutions has too often resulted in factions, jealousy, and rivalry that have done great harm to the testimony of Christ.[57]

We know that these parachurch organizations exist and we know that they tend to pull resources away from the Church, so why do they exist, where did they come from? The concept of the parachurch organization is not biblical as the Church is God's ordained institution in the Bible and ministry flows out from the Church. It is the Church, this gathering of believers who are salt and light in the world. So why did these members of the assemblies of believers suddenly think that it was necessary to create a new institution that is apart from the Church but still connected in some para sort of way? It has been said that most parachurch organizations were formed by entrepreneurs who became frustrated trying to work with the Church and therefore started their own organizations. This simply means that at the

---

[57] William MacDonald, *Parachurch Organization* [article on line] (accessed 17 December 2010) available from http://web.singnet. com.sg/~syeec/literature/parachurch.html

beginning of the parachurch organization there was a time in which a person saw a ministry need and sought to fill that need. He went to the Church and proposed that this need be met from the Church. To do this a person will need to submit to the bureaucracy of the Church. This immediately creates a problem because most Evangelical Churches have a congregational form of government and even when they have a formal board of elders, those Elders are very sensitive to the desires of the congregation. This is all well and good but it takes a great deal of time, argument, debate etc. for a Church to make a decision and then when the decision is reached it carries with it a lot of baggage in terms of conditions, reporting and general meddling. The person making the proposal for the ministry soon abandons the Church and starts his own institution otherwise known as a parachurch organization. The Church feels bad about not doing what they are biblically called to do so they agree to financially support the ministry. In this way they can be associated with a ministry simply by writing a check but without any other commitment. The parachurch organization can succeed or fail without its failure reflecting poorly upon the Church. The parachurch organization can then spread out its base of support across doctrinal and sectarian lines and operate without the bother of Church bureaucracy. Other parachurch organizations did not bother with the Church at all but simply decided that the Church was not doing what was needed to be done and so they went ahead and formed other institutions including Bible Colleges, Seminaries and Mission Agencies. The Church was now relieved of its responsibility to educate and train the Saints for ministry, to recruit and send them off to foreign lands or to do most any ministry in the community what-so-ever.

This leaves us with the question of what can we do about it? Parachurch organizations are so entrenched in the Christian psyche and in the Christian pocket book that they are not going away. The Church cannot afford to ignore them and in fact cannot work without them. For example if a Church were to go to a hospital and offer to leave a Bible in each room, the response would be, the Gideons already do that, we do not need you. If the Church were to send an evangelist to another country the task would be a lot easier if it were done through a mission agency. Another problem is that over the years there is a degree of animosity that has been built up between the Church and the parachurch organizations to the extent that it is difficult for them to work together. The Church has come to resent the parachurch and its freedom of movement without accountability to the Church. The parachurch organization sees the Church as ineffective, lacking creativity or vision and overall incompetent when it comes to ministry whether local or global. However, the parachurch will still take its monthly checks, ask for time in the Church service to present their ministry and make an appeal for even more money. In spite of the animosity the Church will do as the parachurch organization has asked and sooth their conscience by acquiescing to their requests.

Even with a casual reading of the book of Acts it is obvious that the Church, as God's ordained institution, works really well. By the time Paul was finished with the ministry the Lord had given to him there were Churches planted all over Asia minor and these Churches were growing and planting still more Churches to the point where it could be said that "Moreover you see and hear that not only at Ephesus, but throughout almost àll Asia, this Paul has persuaded and turned away many people, saying that they are not gods which are made with hands." (Acts 19:26). Paul could say with confidence, "I have fought the good fight, I have finished the race, I have kept the faith." (II Timothy 4:7). All this work was done within the framework of the Church. Therefore, on the one hand we have the Church as God's ordained institution and the parachurch which exists and will always exist regardless of what anyone may think or say about it. To bring these two together will be the key to advancing of the cause of our Lord in these last days.

The following is a proposed solution to the dilemma which consists of four main components. The **first of these is to recognize the unity of the one body of Christ**. As previously discussed, all who have been born again through faith in Christ are a part of one body whether they choose to work within the framework of the Church or not. Regardless of how impatient and even self-willed a person may be they are still a part of the body of Christ. Regardless of how a person may worship or what denomination they might choose to belong, because of their faith in Christ, they are all a part of that one body. Each member of that one body is to exercise their spiritual gifts and their natural ability to the building up of that one body and to make disciples of the nations. This vision of the one body of Christ composed of many members who are different with different functions and yet one in Christ with one mission and goal ought to have a certain priority in our function as a local Church and as individual believers.

The **second component is that the parachurch organizations must recognize the authority of the Church as God's ordained institution on the earth**. This means that in some very practical manner and at some level the parachurch needs to demonstrate its submission to the Church. For example, there is a parachurch organization which states in their statements that they are an outreach ministry of a specific local Church. In this manner they recognize the authority of the Church and have an accountability connection to the Church. As previously stated, the parachurch organization works primarily in the area of evangelism. In working with the Church, when a person trusts Christ, there is a Church which they can immediately become a part of to worship, serve, and be discipled. The Church can then have an intrinsic role to play in the recruiting and training of volunteers for the parachurch ministry. In this scenario the Church has more than just a passive interest in the ministry i.e. just writing monthly checks but is involved in the actual workings of the ministry.

The **third component is that the Church must step up and take responsibility for reaching its community and the world for Christ**. This includes being involved in ministries which are a service to the community which will cause the community to glorify God (Matthew 5:16) which will provide opportunity for the believer to explain the way of salvation. This includes the recruiting and training or equipping of the Saints to do ministry (Ephesians 4:12). This leaves each local Church with the question of what are we going to do, where do the skill sets and gifts of our Church best fit into which ministry. No one local Church can meet all the needs of a community, let alone reach the entire world for Christ. It will require all these local Churches working together as one body. Therefore, each local Church will need to access their strengths and weaknesses, determine what specific area of ministry for which they are best suited and then do it! The doing of ministry will then open up opportunities for evangelism as seen in previous lessons.

The **fourth component, the last but not the least, is reliance upon the working of the trinity.** We see the trinity at work in creation where the Father formed space and matter and time since space and matter must dwell in time; then we see the Holy Spirit energizing; then the Son spoke it all into existence giving it shape, form, color etc. (Genesis 1). In like matter we see the trinity in salvation as it is the Son who gave His blood through the Spirit to the Father (Hebrews 9:14). It should come as no surprise then when we see the trinity involved in ministry. In Luke 10 we see the 70 sent out by the Son with the understanding that their needs will be provided for by the Father and they will work in the power of the Spirit. We see this all summarized in verse 21 where our Lord prayed, "In that hour Jesus rejoiced in the Spirit and said, I thank You, Father, Lord of heaven and earth, that You have hidden these things from the wise and prudent and revealed them to babes. Even so, Father, for so it seemed good in Your sight. All things have been delivered to Me by My Father, and no one knows who the Son is except the Father, and who the Father is except the Son, and the one to whom the Son wills to reveal Him." Just like the Apostle Paul we need to have confidence in the Lord, the Holy Spirit and the Word of God to do their work in the lives of new believers and in the building of the Church. We must not take too much upon ourselves to think that we are indispensable and thereby negate the work of the Holy Spirit in the lives of new believers where ever they may be located. This will free us up to move on to other locations and to the many others who need to have the opportunity to hear, understand and believe the Gospel.

As stated previously many parachurch organizations began with a believer who is frustrated with the bureaucracy of the Church which is based in congregational rule. Inasmuch as leadership is an important function of any organization it will be good to examine biblical leadership. If the local Church had a biblical form of government and leadership it would be a big encouragement to those ministry entrepreneurs who want to see things get done to advance the cause of our Lord Jesus Christ. The following paragraphs are taken from my doctoral dissertation titled *The Value of the New Testament Church in the Post Modern World* and will be footnoted at the end of this segment.

There are basically two types of government practiced in this day and age by the Church. The one is known as congregational rule and the other is known as elder rule. It can be argued that American democracy came as a result of congregational rule. It can also be argued that congregational rule came as a result of the principles of democracy which originated in Athens during the golden age of Greece. Elder rule is found in the Scriptures, therefore a study of the Scriptures is important when seeking to understand Church government.

In the book of Acts during the travels of Paul and Barnabas they are seen preaching the gospel and then as the believers came together as the Church, they ordained elders in every Church Acts 14:23. This word comes from the Greek word presbuterous or *presbuterous*. The ordinary usage of the word is that of an older man, one advanced in years. In the Scriptures the word came to denote those who were to lead the Church or what has been called the *presbytery*. The first thing that is established here is that elders are older or more mature men. They are older in this life and they are mature in spiritual things. The context will tell if the word is being used to denote an old man or a leader of the Church. Another word that is used to describe these men is the word επισκοπηs or *episkopas*. This word is literally *upon see* or overseer. It is also translated Bishop. This word has more to do with what they do i.e. they oversee the flock as a shepherd. We have the word *presbuterous* to tell who they are i.e. older in the faith and the word *episkopas* to tell what they

197

do i.e. oversee the flock to be sure they are protected. Both of these words have to do with the same person as Alex Strauch points out, "A number of New Testament passages make it obvious that the two terms refer to one and the same group and are used interchangeably:"[58]

The words used to describe these men are always in the masculine gender and are always plural. The idea that one man, who is the "pastor," is the only elder is not taught in the Scriptures. The Scriptures teach that the Apostles appointed elders Acts 14:23; they were chosen by Apostolic legate Titus 1:15; they are set apart by the Holy Spirit Acts 20:28; and they are known by their works I Timothy 3:1. In I Timothy 3:1 we read, "This is a faithful saying, if a man desires the position of a bishop, he desires a good work." The desire for the work is what sets a man apart as an elder which is how the Church knows that the Holy Spirit has made him an elder by the work that he does on behalf of the Lord's people. The desire for and the doing of the work should be the chief concern of the Church for their elders.

This leads to the question, what do they do? What is the work of the elder, this work that they are to have a strong desire for and to be known for? Again, the Scriptures are clear as to what we should expect to see

---

58   Alexander Strauch, *Biblical Eldership* (Littleton: Lewis and Roth Publishers, 1986) 205.

from our elders. In Acts 20:28 we read, "Therefore take heed to yourselves, and to all the flock, among which the Holy Ghost has made you overseers, to shepherd the church of God, which He purchased with his own blood." The work of an elder is given in terms of a shepherd and his flock. A shepherd will take heed of his flock and will keep watch over them. The image here is of a shepherd standing on a high hill and looking over the flock so that he can see if they are well provided for and if they are safe. The Holy Spirit has set them aside for this work and one of the ways in which the Church knows that the Holy Spirit has set them aside for the work is that they are on the hill looking after their welfare. They are to feed the flock and this flock is very special because it has been purchased by the blood of our Lord, the chief shepherd. It is to be argued here that to feed the flock is to teach them the Word of God. The Scriptures are our daily bread, as our Lord referred to Himself as the bread of life, John 6:35,48, the way in which we feed upon Him, John 6:53-58, is by the study of the Bible and the teaching of the Holy Spirit.

In I Peter 5:1-3 we read, "The elders who are among you I exhort, I who am a fellow elder, and a witness of the sufferings of Christ, and also a partaker of the glory that will be revealed: Shepherd the flock of God which is among you, serving as overseers, not by compulsion but willingly; not for dishonest gain, but eagerly; nor as being lords over those entrusted to you, but being examples to the flock." Once again the work of an elder is that of a shepherd feeding the flock and watching over them. In this text the elders are exhorted to do this work with a willing spirit, not for the sake of money and they are not to be lords over the flock but to lead by example. Unfortunately there are those who mistakenly view the instruction of not being lords over them as a rational for passive leadership. In fact there is no such thing as passive leadership as this is not leadership. What happens is that those whom the Lord has placed into leadership abrogate the responsibility of leadership over to congregational rule. This is where decisions are made by way of committee and the vote of the people. It is interesting that the Greek word Laodicea, the name of the seventh Church in Revelation 3, literally means the "rights of the people". As Harry Ironside so aptly states:

> Laodicea is a compound word, and means "the rights of the people." Could any other term more aptly set forth the condition of present-day church affairs? It is the era of democratization, both in the world and in the church. The masses of the people are realizing their power as never before. The terrific slogan, vox populi, vox Dei (The voice of the people is the voice of God), is ringing through the world with clarion-like distinctness.[59]

---

[59] H. A. Ironside, *Lectures on the book of Revelation* (New Jersey: Loizeaux Brothers, 1975, reprint) 74.

The first problem with congregational rule is that which is described in Revelation 3:15-17, "I know your works, that you are neither cold nor hot. I could wish you were cold or hot. So then, because you are lukewarm, and neither cold nor hot, I will vomit you out of my mouth: Because you say, 'I am rich, have become wealthy, and have need of nothing—and do not know that you are wretched, miserable, poor, blind, and naked—" These Christians who are lukewarm to spiritual things but are rich, increased with goods and in need of nothing are now expected to vote on the spiritual direction of the Church. The second problem is closely related to the first in that it is quite common for there to be a majority of carnal Christians in a Church. Therefore, any vote that is taken will invariably lead to more carnality to the point that the desires of the carnal Christian will trump the true will of God. This is seen in the Church of Corinth where there was sexual immorality and rather than mourn the Corinthians were proud of their tolerance and grace to this man. The Apostle Paul set them strait in instructing them to deliver this one to Satan and to purge out this leaven (sin) I Corinthians 5:5,7. Give carnal Christians the vote and sin and rebellion are the end result.

Therefore the Lord has set spiritual men to lead, feed and care for His flock. The elders are to watch over their souls, Hebrews 13:17; they are to labor and admonish,

I Thessalonians 5:12; they are to rule well, I Timothy 5:17; Hebrews 13:7; they are to be good stewards of God, Titus 1:7. The response of the Church to these elders is to know them, I Thessalonians 5:12; esteem them very highly, I Thessalonians 5:13; trust them, I Timothy 5:17-22; remember them, Hebrews 13:7; obey them, Hebrews 13:17; greet them, Hebrews 13:24; pray for them, Hebrews 13:18.

The fear of those who hold to congregational rule is that these men will become dictatorial or will lord it over them. This would be true under two assumptions. The first is that they have not been set aside by the Holy Spirit and are not shepherds but thieves or robbers. The second is that they do not meet the qualifications of an elder. The qualifications of an elder are found in I Timothy 3:1-7 and Titus 1:5-9. If one were to collect the qualifications listed in both of these texts it would look something like this:

POSITIVE	NEGATIVE
Vigilant	Not given to wine
Sober	Not a striker
Apt to teach	Not greedy
Patient	Not contentious
Lover of good men	Not a novice
Holy	Not coveteous
Temperate	Not a brawler
Hold to the Word	Not double tongued
Good behavior	Not self willed
Good report	Not slanderous
Blameless	

Just
Husband of one wife
Show hospitality
Rule well his own house
Have his children in subjection

Any man who meets these qualifications and has been set aside by the Holy Spirit can be depended upon to serve the Lord's people as a servant leader and as a shepherd who cares for the Lord's flock. This group of godly men, being led by the Holy Spirit, are to study the Scriptures and pray to determine the Lord's will and then lead the people by their example while feeding them the Word of God. This does not mean that they will never consult with other members of the Church, particularly those who have certain gifts and expertise in the area under consideration, they certainly will avail themselves of all the gifts given by the Lord to the Church. However, elder rule is the biblical form of leadership that comes down from our Lord who is the head of the Church.[60] With respect to evangelism, this form of government will serve to expand the ministry of the Church to reach more people for Christ. This form of government will encourage those who wish to launch out into ministry while at the same time have the accountability and protection of the local Church.

---

[60] Greg Koehn "The Value of the New Testament Church in a Postmodern World" (PhD diss., Louisiana Baptist University, 2013) 35-40

Therefore, let us pray that each of us individually will see the value and the beauty of the local Church and commit ourselves to it with our time, money and abilities as unto the Lord. For the Church is God's ordained institution on earth and is the bride of Christ in glory, ". . . Christ also loved the church and gave Himself for her, that He might sanctify and cleanse her with the washing of water by the word, that He might present her to Himself a glorious church, not having spot or wrinkle or any such thing, but that she should be holy and without blemish." (Ephesians 5:25-26). "Let us be glad and rejoice and give Him glory, for the marriage of the Lamb has come, and His wife has made herself ready. And to her it was granted to be arrayed in fine linen, clean and bright, for the fine linen is the righteous acts of the saints." (Revelation 19:7-8). And may this glorious Church step up and take her rightful place in local and global evangelism.

Read More About It

Weldon Adams, *The Church According To The Scripture*
W.T.P. Wolston, *The Church: What is it?*
Alexander Strauch, *Biblical Eldership*

CHAPTER TEN

THE KINGDOM AND THE CHURCH

We hear and read a lot about the kingdom. It has gotten to the point where people will speak more about the kingdom than they do the Church. Just recently I had a conversation with a fellow believer who was referring to someone else as one who advanced the kingdom. I am going to start stopping people and asking what they mean by that? Are we to be building the kingdom or are we a part of what God is doing to build His Church? Whatever happened to the Church? What is wrong with the Church anyway?

I realize that there are those who would reply that there is a great deal wrong with the Church, books have been written where there are chapters upon chapters criticizing the Church. There are at least two problems with this criticism. First, as noted in the previous chapter, the Church is the only institution ordained by God. Our Lord stated that He will build His Church (Matthew 16:18). The Church is intimately linked to the body of

Christ, "And He is the head of the body, the Church." (Colossians 1:18). The fullness of God is found in the Church, "And He put all things under His feet, and gave Him to be head over all things to the Church, which is His body, the fullness of Him who fills all in all." (Ephesians 1:22-23). To criticize the Church is to criticize the person and work of our Lord Jesus Christ. Second, the complaints being made against the Church are being made against the people who are responsible for the practice and behavior of the Church i.e. all believers. Therefore, in some sort of technical sense the criticism is not directed at our Lord but at His people and their behavior. After all the Church is not a building it is the gathering of believers. But since these believers are all members of the body of Christ should we condone this criticism? Perhaps we can suffice it to say that the criticism is of the actions or lack of acceptable actions of believers rather than believers themselves. Enough of that, it is time to get back to the kingdom.

It has been said that there are 155 references to the kingdom in the New Testament and this is true. The Gospels contain the largest use of the word with 124 references, of those 54 are found in the Gospel of Matthew and 44 are found in the Gospel of Luke. Of course in the Gospels many of the references are repetitious particularly in the synoptic Gospels. Nevertheless, there is a large number and we should take the subject of the kingdom seriously. It is the Gospel of Matthew where the kingdom is of major importance because the theme of the Gospel is that the Lord Jesus is King and has come to announce and make an offer of His kingdom to Israel.

In the Gospel of Matthew we see at the beginning of our Lord's ministry, He went out to preach and to say, "Repent, for the kingdom of heaven is at hand." (Matthew 4:17). Repentance is a particularly Jewish expression and command. Israel is God's chosen earthly people, they were the first to receive the covenants and promises of God (see Romans 9:4). They persisted in departing from God and needed to be told to return or turn around from the direction they were taking and return to their God, hence to repent. In Matthew 4:23 we read, "And Jesus went about all Galilee, teaching in their synagogues, preaching the **gospel of the kingdom**," Here it will be important to understand this gospel. We need to think about what is the gospel of the kingdom and ask the question, is it the same gospel that we preach for the Church?

In Matthew 4:23 we read, "And Jesus went about all Galilee, teaching in their synagogues, preaching the **gospel of the kingdom,** and healing all kinds of sickness and all kinds of disease among the people." Some of the things that we consider when we exegete a passage of

Scripture is grammar, immediate context and the context of all Scripture.

The grammar of the text is very straight forward, the word *gospel* is the same as is used throughout Scripture. It is translated from ευαγγελιον which is the same word used in I Corinthians 15:1. The word means good news. So in the Gospel of Matthew we have the *good news* of the kingdom and in I Corinthians 15 we have the *good news* that Christ died for our sins and rose from the dead. The word *kingdom* is translated from βασιλειας (*basileia*) which has to do with the reign of a king i.e. His authority and rule. J.D. Pentecost in his book *Things to Come* quotes from Ladd who gives an excellent explanation of the word:

> The primary meaning of the New Testament word for kingdom, *basileia*, is "reign" rather than "realm" or "people." A great deal of attention in recent years has been devoted by critical scholars to this subject, and there is a practically unanimous agreement that "regal power, authority" is more basic to *basileia* than "realm" or "people". "In the general linguistic usage, it is to be noted that the word *basileia*, which we usually translate by *realm, kingdom,* designates first of all the *existence,* the *character,* the *position* of the king. Since it concerns a king, we would best speak of his *majesty,* his *authority*" (Schmidt, *Theologisches Worterbuch zum Neuen Testament,* I, p. 579)[61]

---

[61]   J. Dwight Pentecost, *Things To Come* (Grand Rapids: Zondervan Publishing, 1958, reprint 1971) 429-430.

Therefore, it is important for us to understand that the kingdom being presented has to do with a king who has regal power and authority. Remember the theme of the Gospel of Matthew is the presentation of a king who is our Lord Jesus Christ.

The manner in which the King proves His authority is by a demonstration of His power. This is why when our Lord preached the gospel of the kingdom, He exercised His power by "healing all kinds of sickness and all kinds of disease among the people." This is consistent with what the prophets foretold concerning the Messiah, "Then the eyes of the blind shall be opened, and the ears of the deaf shall be unstopped. Then the lame shall leap like a deer, and the tongue of the dumb sing." (Isaiah 35:5-6). When John the Baptist sent the question to Him as to if He were the one, His response was, "The blind see and the lame walk; the lepers are cleansed and the deaf hear; the dead are raised up and the poor have the gospel preached to them." (Matthew 11:5). Simply stated, the gospel of the kingdom is that the King has come to establish His kingdom and He verifies His message with demonstrations of His power.

The concept of the kingdom is seen throughout all of Scripture. Therefore, it is important to understand this concept in the context of both the Old and New Testaments. I will not clutter up space by noting all the various views of the kingdom and will attempt at not becoming too tedious in the examination of Scriptures but rather give a synopsis of what is taught. J Dwight Pentecost does an excellent job of examining all the aspects of the kingdom in his book *Things To Come* particularly in the section beginning with page 427 to 475 for those who want the tedious version.

The concept of the kingdom is not something new introduced by our Lord in the Gospels. It is very much a part of prophecy and God's dealing with mankind and the nation of Israel in particular. Prophecy concerning the kingdom is one of a mixed message. The kingdom is seen as spiritual and physical; it is seen as the present or near future and as the distant future; it is seen as being for a fixed time and as being an eternal kingdom; it is seen as pertaining to Israel and to the entire Gentile world. Therefore when confronted with the question of which is it? The answer is both/and.

That the kingdom is spiritual is seen in the words of our Lord that for a person to see or enter into the kingdom they must be born again (John 3:3,5). This new birth is by the Holy Spirit upon believing faith in Christ. Our Lord refers to the kingdom as a mystery in Matthew 13. The parables given in that text have a spiritual meaning as our Lord Himself explains. At the same time the prophets saw a real physical kingdom. The teaching of this kingdom goes back to the Abrahamic covenant. When the Lord called Abraham out to go to a place that He would show him He said, "To your descendants I will give this land." (Genesis 12:7). Then the Lord told Abraham that his descendants shall be as the stars in the heaven without number (Genesis 15:5). God established His covenant with Abraham in stating, "And I will establish My covenant between Me and you and your descendants after you in their generations, for an everlasting covenant, to be God to you and your descendants after you. Also I give to you and your descendants after you the land in which you are a stranger, all the land of Canaan, as an everlasting possession; and I will be their God." (Genesis 17:7-8). There are at least three components to this covenant as seen in this text. The first is that God will be the God of Abraham and his descendants. This established the supreme rule of God over the nation of Israel. The second component is that Abraham's descendants will be without number and the third component is that these people will occupy the land that Abraham now traveled. The fact that a physical people will occupy a physical land speaks to a physical kingdom. This covenant has one more component found in Genesis 22:17-18 "blessing I will bless you, and multiplying I will multiply your descendents as the stars of the heaven and as the sand which is on the seashore; and your descendents shall possess the gate of their enemies. In your seed all the nations of the earth shall be blessed, because you have obeyed My voice." This fourth component is that Abraham's people will be blessed and will be a blessing to all other nations. Ultimately the greatest of blessing is found in the person and work of our Lord Jesus. At the same time there are a number of ways that Israel has and continues to bless the nations through their intelligence and ingenuity. It is important to keep in mind that the Abrahamic covenant is unconditional, it is that which God will do apart from human actions.

This physical kingdom is to be ruled by a literal, physical king who will come from the line of David. This is known as the Davidic covenant which is also an unconditional covenant as we read, "And your house and your kingdom shall be established forever before you. Your throne shall be established forever." (II Samuel 7:16). From this we understand that the king (Messiah) will come from the line of David. Hence in Matthew our Lord's linage is traced through Joseph back to David and in Luke His linage is traced through Mary to David.

Pentecost describes the spiritual character of the kingdom as consisting of *righteousness, obedience to the will of God, holiness, truth,* and the *fullness of the Holy Spirit.* These are taken from both the Old Testament and the New Testament.[62] In the Old Testament both the major

---

[62]   Ibid., 482-486

and minor prophets spoke a great deal about the kingdom, the following are conditions existing within the kingdom:

- *Peace*
- *Joy*
- *Holiness*
- *Glory*
- *Comfort*
- *Justice*
- *Full knowledge*
- *Instruction*
- *The removal of the curse*
- *Sickness removed*
- *Healing of the deformed*
- *Protection*
- *Freedom from oppression*
- *No immaturity*
- *Reproduction by the living peoples*
- *Labor*
- *Economic prosperity*
- *Increase in light*
- *Unified language*
- *Unified Worship*
- *The manifest presence of God*
- *The fullness of the Spirit*
- *The perpetuity of the millennial state*

Even with this very cursory view of the kingdom, it ought to be obvious that this is not something that can or will be brought about by human effort. The kingdom must be established by God and God alone. Any attempts by mankind to establish this kingdom is human arrogance which will fall far short of what God intends in His own time. This kingdom is earthly and it was first presented to the Jew. The city of Jerusalem will be the capital and Israel will be at the center of the Messiah's reign, "And in that day His feet will stand on the Mount of Olives, Which faces Jerusalem on the east" (Zechariah 14:4); "The people shall dwell in it; and no longer shall there be utter destruction, But Jerusalem shall be safely inhabited." (14:11); "And it shall come to pass that everyone who is left of all the nations which came against Jerusalem shall go up from year to year to worship the King, the Lord of hosts, and to keep the Feast of Tabernacles." (14:16).

It is to be noted that the kingdom is also referred to as the millennial kingdom. The reason for this is found in the Revelation:

> He laid hold of the dragon, that serpent of old, who is the Devil and Satan, and bound him for a *thousand years*; and he cast him into the bottomless pit, and shut him up, and set a seal on him, so that he should deceive the nations no more till the *thousand years* were finished. But after these things he must be released for a little

while. And I saw thrones, and they sat on them, and judgment was committed to them. Then I saw the souls of those who had been beheaded for their witness to Jesus and for the word of God, who had not worshiped the beast or his image, and had not received his mark on their foreheads or on their hands. And they lived and reigned with Christ for a *thousand years*.

Now when the *thousand years* have expired, Satan will be released from his prison. (20:2-7)

The sum of what we have thus far is that the kingdom envisioned by the prophets is a literal, physical kingdom here on earth, primarily for the Jews but will affect all the peoples of the world and will be ruled by our Lord as king and as the Jews' Messiah sometime in the future. This kingdom is temporary in the sense that there will be a one thousand year reign of Christ and it is eternal in that the reign of our Lord will then continue throughout eternity.

As mentioned earlier the theme of the Gospel of Matthew is that our Lord is presenting Himself as the King or Messiah of Israel. Our Lord makes a legitimate offer to the nation of Israel of Himself as their King who will then rule over His kingdom. The problem is that the Jews rejected their King and His kingdom. A cursory view of the gospel of Matthew will now be undertaken to help us understand this kingdom gospel. Pentecost summarizes the book of Matthew in the following manner: "There are three major movements in the Gospel of Matthew: (1) the presentation and authentication of the king (1:1-11:1); (2) the opposition to the King (11:2-16:12); and (3) the final rejection of the King (16:13-28:20)."[63]

In the **first movement** (1:1-11:1) we have the birth of the King where all the prophetic requirements for the birth of the Messiah are met and described. The generations of our Lord are given to prove that He has a legal right to the throne of David. At His birth lowly shepherds and Gentiles come to worship Him while the king of the Jews tries to kill Him. The ministry of John the Baptist is described and so meets the requirements of the prophesied forerunner. His temptation proves that He is morally qualified to be the King. He demonstrates His regal authority in bringing men into obedience to Him and His power as He works miracles of healing. In His sermon on the mount He gives "the requirements for entrance into this anticipated kingdom."[64] It could be said that the Sermon on the Mount gives an overview of what a society would look like under His rule. He continues to prove His power and authority in His healing of all manner of diseases, cast out demons, forgives sins and raises the dead. He commissioned His disciples to go

---

[63]   Ibid., 456.
[64]   Ibid., 457

forth and explained the cost and compensation of what it means to be His disciples.

In the **second movement** (11:2-16:12) we see the development of the rejection of the King. First His forerunner John is rejected and then He announces woe upon the cities where He worked and yet they did not repent. At the end of chapter 11, He gives an individual invitation. His attention is turning from the nation of Israel to individual men and women who will come to Him, "Come to Me, all you who labor and are heavy laden, and I will give you rest." (11:28). Chapter 12 of Matthew is a key section of the narrative because it is the turning point at which our Lord is rejected by the Jews and now the gospel to the Gentiles is intimated. To put it another way the *mystery* part of the kingdom is about to be introduced. In verse 14 we read, "Then the Pharisees went out and plotted against Him, how they might destroy Him." Our Lord then quoted from Isaiah 42:1-4, we want to note verse 21 in particular, "And in His name Gentiles will trust." The Pharisees go on to ascribe the work of the Holy Spirit to Satan and later ask the Lord for a sign that they might believe. Our Lord responded, "An evil and adulterous generation seeks after a sign," (vs 39). The rejection of the King and His Kingdom is fast approaching its final movement.

In chapter 13 we have what is known as the mystery of the kingdom. It is interesting that Matthew refers to these parables as the mystery of the kingdom of heaven while Mark and Luke refer to them as the mystery of the kingdom of God. Other authors have tried to make something out of the difference but basically they are all talking about the same thing. The one possible difference is the perspective and audience of the three writers. The Jewish audience of Matthew could be offended by referring to God while understanding exactly what is meant by heaven or that which is of heaven. In any event the *mystery* refers primarily to the Church. Whenever the word is used in the Epistles it invariably refers to that which is a part of Church doctrine. By definition the *mystery* is that which was previously concealed but is now revealed. Having been rejected by the nation of Israel our Lord now turns His attention to this mystery which is the Church. He begins this line of teaching with parables which teach in very broad strokes what the Church age will look like. It is very important to keep in mind that this teaching is broad and even generic. One must use caution to make too specific an interpretation or even application of the meaning of these parables.

A summary of these parables is now submitted to help in understanding the mystery or spiritual form of the kingdom. The period of this mystery is what I like to call the hidden valley in the mountain peaks of prophecy. The mountain peaks of prophecy are that the prophets saw the coming of the Messiah and the cross; then they saw the King and His kingdom. They did not see the mystery that lies in the valley between the peaks of the cross and the kingdom. In the parable of the sower and the soils we see that the Word of God will be preached during this time. There will be four different types of hearers of the Word who will be influenced by the devil, the world and the flesh. This means that out of the four hearers of the Word only one will actually take it in and bear fruit as a result. The fruit will vary in its production, some a hundredfold, some sixty, some thirty. Therefore, as we preach and teach the Word of God we expect that it will be rejected by many and received by few. Among those who receive it there will be different levels of spiritual maturity. It is interesting that a sovereign God does not force everyone to receive the Word and bear an abundance of fruit. Instead He allows humans to exercise their will and chose to take in His Word and then allows them to produce whatever fruit they chose to produce. The only possible reason for this is that God desires to be loved by His creation mankind and love is defined by our choices to obey Him.

The second parable is similar to the first one in that there is the sowing of seed. In this one we learn that there will be a sowing of good seed and of bad seed. It is easy to see how that during this time the Word of God is being proclaimed while at the same time there is a tremendous amount false teaching going forward. Much of the false teaching is disguised as the true teaching. There are many who call themselves Christians and yet do not know for sure if they are going to heaven nor do they know anything of justification by faith and what it is to be born again. The only way to know the difference is by their fruit. The good seed will bring forth good fruit like seed that brings forth wheat. The bad seed will produce weeds. This parable is projected to the end of time where the bad will be taken out and burned and the good which has been left will shine in the kingdom. This is not the rapture as it is the bad that is taken first and then the good. The final judgment will then separate the bad from the good and the good will move forward into the eternal kingdom.

The third parable is of a mustard seed that becomes a great tree. There are those who teach that this is the Church that begins small but grows to be very large. It is important to note that the natural growth of the mustard seed is to become a bush, not a tree. The birds of the air rest in the branches of this great tree and birds have already been identified with Satan in a previous parable (vs 19). It is true that what we have today is that which is a monster in arrogance and pretention and the servants of Satan are welcome and very comfortable. So that if the mustard plant represents the Church it is representing the false church that comes from a false gospel.

The fourth parable is that of leaven being placed in three measures of meal by a woman. However we may think of this parable we must keep in mind that leaven invariably speaks of sin. It is as our Lord taught His disciples to beware of the leaven of the Pharisees and the Sadducees (Mt. 16:6) and the leaven of Herod (Mk. 815). In each text leaven represents sin which could be the sins of legalism, skepticism and materialism. In I Corinthians 5 we have instructions for Church discipline which is taught in terms of purging out the leaven, "Your glorying is not good. Do you not know that a little leaven leavens the whole lump? Therefore purge out the old leaven, that you may be a new lump, since you truly are unleavened. For indeed Christ, our Passover, was sacrificed for us." (I Corinthians 5:6,7).

The fifth parable is of a treasure hid in a field and a man sells all that he has and buys the field. It is interesting that in Exodus 19:5 we read, "Now therefore, if you will indeed obey My voice and keep My covenant, then you shall be a special treasure to Me above all people; for all the earth is Mine." It is certainly true that all the earth belongs to the Lord especially in view of His perfect sacrifice where He gave His all. It also fits the concept of the mystery being the Church age where Israel has been set aside for the moment until the Lord is ready to bring them forth again.

The sixth parable is a bit different in that in this parable the man bought the pearl of great price rather than buying the field as in the last parable. A pearl is formed by gradual accretion and is used as an adornment. Certainly it can be said of the Church that it is being formed by gradual accretion as souls are being added to the Church daily. Then one day the Church will be caught up to be with the Lord to be His bride and His special adornment. This paints a picture of the Church has being within the framework of the kingdom. It is like the Church is a circle within the larger circle of the kingdom in the spiritual sense.

The seventh parable is a summary of the end of the age. The kingdom of heaven is like a drag net that is cast into the sea and when the net is pulled in the fish will be separated out where the good is kept and the bad is thrown away. This and the other parables do not state what it is that makes the bad, bad and the good, good. We have to continue reading and go to the epistles especially the book of Romans where we learn that there are none good but that through faith in Christ one can be justified and accounted as good.

Then there is an eighth and final parable. The kingdom of heaven is like a householder who brings out of his treasure things new and old. There are a number of meanings attached to this which include that it refers to the old and new testament. The problem with that is no one at that time would have known that there is such a thing as a New Testament. That would be okay in the sense that we are dealing with a mystery. Another view is that it is good to have both the old and the new. Christians of the this day and age could certainly learn from that because there is a tendency to grab onto the latest new thing and run with it at the expense of the tried and true traditions of the Church or to put it another way, to throw the baby out with the bath water. It is true that we hold onto our doctrine but have permission to change our method. The caution is that many times our method will reflect our doctrine. It is also true that there are many old traditions that need to be thrown out, but at the same time keep in mind that there are many old traditions which have withstood the test of time for a very good reason.

We may now summarize the parables in this manner. The King has offered the nation of Israel His rule and Kingdom. The nation is in the process of rejecting Him and His rule. This brings us into a period of mystery. This mystery period is presented in very broad strokes leaving room for the details to be filled in later on in the Epistles. Even though Israel as a nation has entered into this movement toward the ultimate rejection of our Lord, the priority of our Lord's message is still to Israel and to fulfill the Abrahamic and Davidic covenants The ultimate rejection of our Lord by Israel will be when they declare that they will not have this man to rule over them and He is nailed to a cross. The broad strokes of the mystery period are: 1) the Word of God will be preached and rejected by a majority of people but there will be some who will embrace the Word as true and believe; 2) there will be false gospels preached and it will not be possible to tell the difference except by the way in which the people live or by their fruit; 3) there will be an unnatural growth of what seems to be the truth but the servants of evil will find a place in their midst; 4) there will be sin in the mix of the true believers so that they will need to be on their guard to put out the sin from their midst; 5) there will be two entities in the kingdom, a treasure and a pearl where the Lord will purchase the whole field to have the treasure and then the pearl of great price; 7) the age will end with a final sorting out between the good and evil; 8) it is good to have a combination of both old and new things.

At this point it is important to keep before us how the emphasis is on the Word of God and of the sovereign working of God in procuring His prize possessions and determining the good from the evil. At this point there is no discussion of doing good works or of making this world a better place to live for all humanity. There is no indication that mankind is capable of creating or establishing the kingdom on its own.

Another section of major importance concerning the kingdom is found in Matthew chapters 24 to chapter 26. This is known as the "Olivet Discourse". It is important to understand that this discourse is given in answer to the questions of the disciples, ". . . when will these things be? And what will be the sign of Your coming, and of the end of the age?" (Matthew 24:3). Clearly this discourse has to do with the end of the age, at the coming of our Lord to set up His kingdom. This is emphasized again in verse 15 where we read, "Therefore when you see the abomination of desolation, spoken of by Daniel the prophet, standing in the holy place". This reinforces two things, first the abomination set up by the Greeks during the Greek and Persian wars around 400 to 100 BC was not a fulfillment of prophecy. Second, this is an event which is to take place after the Old Testament sacrificial system has been reinstituted, in the end times. In the verses and chapters that follow the great tribulation is described and watchfulness is enjoined. This discourse is about what happens during the tribulation period and has to do with the Jews. They who rejected their Messiah at

His first advent will suffer the consequences during the tribulation, the time of Jacob's trouble (Jeremiah 30:7) and Daniel's 70th week (Daniel 9:24-27). Then our Lord will return and set up His literal, physical kingdom and the Jews will look upon Him whom they pierced and mourn (Zechariah 12:10).

This brings us to the qualifications for entering into this kingdom. In Matthew 25 verses 31 to 46 we have a very well known text. It begins with when the Son of Man comes in His glory He will gather the nations together and then will divide them as a shepherd would divide his sheep from the goats. The sheep will go on His right hand and the goats on His left hand. Those on His right hand will inherit the kingdom prepared for them from the foundation of the world. Those on His left hand will be sent into the everlasting fire. What is significant here is the basis for being on either the right hand or the left hand. Those on the right hand are those who fed the hungry, gave a drink to the thirsty, were hospitable to the stranger, clothed the naked, visited the sick, and visited those in prison. The fact that they did this to the least of these, they did it to the Lord. Clearly this is a gospel of works, people are granted entrance into the kingdom based upon what they did. Now, is this the gospel that we preach today? Obviously the answer is no. The gospel that we preach today is to believe on the Lord Jesus and you will be saved. During the tribulation period people will be saved by faith but their faith will be proved by what they do such as refuse the mark of the beast and treat others well especially the Jews. This will be like salvation in the Old Testament which relied upon what a person did, there was no sealing of the Holy Spirit as is the case with the Church. We have Old Testament Saints, Church Saints and Tribulation Saints, these in Matthew 25 are Tribulation Saints.

It might be argued that since this text has to do with the tribulation then can it be used today in the Church as a reason or defense of certain ministries? It has been well said that there is one interpretation and many applications of a text. Certainly all Scripture is given by inspiration of God and is profitable, therefore, this passage is profitable for our learning and maturity in Christ. Christians today will want to feed the hungry, shelter the homeless, minister to those in prison etc. However, the motivation and power to do such things will come from the Holy Spirit within us as we are brought to spiritual maturity. In other words we do not do these or any other works to become a Christian or to enter into the kingdom, we do them because we are believers and disciples of the Lord Jesus Christ.

The implications of this gospel of the kingdom and how the Lord's people work this out in their lives is of vital importance. This is where we find the division between the premillennial view; the amillennial view and the postmillennial view. Before defining what these are, it would be helpful to explain dispensationalism. The word *dispensation* comes from the word οικονομια (oikonomia) which is a household or stewardship. It is a basis of exchange like the US economy is based upon the dollar so we could say that we have a dispensation of the dollar, you give a person a dollar(s) and they give you goods and/or services in return. The traditional view of dispensations is that there are seven to eight dispensations in the Bible, each one divided by a major event in human history. There are other forms of dispensationalism but that study is outside of the parameters of this volume. It is important to keep in mind that this is not a doctrine but rather a hermeneutic or a model for interpreting Scripture.

The first dispensation is that of **Innocence** in the garden of Eden. God created man free from sin and evil in innocence. God then put man to the test by commanding that they not eat of the one tree in the midst of the garden. God gave to man innocence and man failed by disobeying God and eating of the forbidden fruit. This resulted in the Fall, the first major event in human history. The second dispensation is that of **Conscience**, let your conscience be your guide. Once again man failed as the Lord looked down upon them and saw that the thoughts and intents of their heart were evil continually. The solution to the problem of evil was to destroy all flesh that breathed air. But Noah found grace in the eyes of the Lord and Noah along with his wife, his 3 sons and their wives were spared the flood in an Ark. The animal species that required air to breathe were also spared. The flood was the second major event. The third dispensation is that of **Human Government.** This is based upon the Noahic covenant where God promised that He would not destroy all life with a flood and that human life was to be held sacred. Therefore, whoever sheds man's blood his blood will be shed for the life of a man is in the blood, this is otherwise known as capital punishment. But once again mankind failed in that they gathered on the plain of Shinar and there they sought to build a city and tower where the top would reach up into the heavens so that they might worship the sun, moon and stars. In other words they would worship creation rather than the creator. God then came down and gave them all different languages so that they would divide up into separate people groups and then spread out over the face of the world. The third major event is the tower of Babel. The fourth dispensation is that of the **Family**. Abraham is called by God out of the Ur of the Chaldeans to go to a place that He will show him. God gave Abraham an unconditional covenant that consists of Land, Seed, and Blessing. Abraham and his descendants would own the land where Abraham was then pitching his tent and his descendants would be a blessing to the peoples of the world with the ultimate blessing being in the person of our Lord Jesus Christ. The fourth major event is that of Mount Sinai. The children of Israel had gone down into Egypt where they were enslaved. The Lord delivered them from their slavery using Moses as His instrument. This led to the fifth dispensation of **Law**. The children of Israel told God that they would do whatever He said to do (Exodus 19:8) so the Lord gave them what they were to do in the law. One of the things that the law has proved is that there is no one who can keep the law, not even the ten commandments, let alone the rest of it found in Leviticus and Deuteronomy. Therefore, we read that ". . . by the works of the law no flesh shall be justified." (Galatians 2:16). This dispensation came to a close with the crucifixion of our Lord, the fifth major event in human history.

Up to this point a chart of the dispensations would look like this:

	Fall		Flood		Tower of Babel		Mt. Sinai		Cross	
Innocence		Conscience		Human Gov't		Family		Law		Grace

The sixth dispensation is the dispensation of **Grace**. This is our present dispensation, it is where God is offering to mankind His grace. As we have we already noted this offer is being rejected by the majority of mankind. In terms of where we are in prophecy this is the mystery period. The Apostle Paul spoke of how the mystery had been made known to him and that now it was his responsibility to make this mystery known (see Ephesians 3:1-12). This is what I call the Mountain Peaks of Prophecy where the prophets saw the coming of the Messiah and then the kingdom but did not see the valley of the mystery, the Church. When Israel rejected their king and His kingdom they were set aside as a nation. The focus is now on the Church and the gospel of our Lord Jesus that whoever believes on Him will be saved. The next major event in the history of mankind will be the rapture.

A few comments concerning a pre-tribulation rapture are in order here. The basic argument for the pre-tribulation rapture rests upon the biblical distinction between the Church and Israel. As already noted the Church is a mystery that was hidden in the Old Testament and then revealed in the New Testament. God has set aside national Israel during the time of the Church. The rapture will remove the Church from the earth and then the Lord will turn His attention back to Israel and to an unbelieving world. This is why the tribulation is called the time of Jacob's trouble (Jeremiah 30:7). The entire seven years of tribulation is a time of the outpouring of the wrath of God upon the world, including Israel (Revelation 6:16-17). The Church does not come under the wrath of God because we are in Christ and we are as He is in this world, "Love has been perfected among us in this: that we may have boldness in the day of judgment; because as He is, so are we in this world (I John 4:17). The wrath of God will never again pour out upon our Lord. In I Thessalonians 4 and verses 13-18, we have an explanation of the rapture and especially in verses 16, 17 and 18 where we read, "For the Lord Himself will descend from heaven with a shout, with the voice of an archangel, and with the trumpet of God. And the dead in Christ will rise first. Then we who are alive and remain shall be caught up together with them in the clouds to meet the Lord in the air. And thus we shall always be with the Lord. Therefore comfort one another with these words." This message of rapture is one of comfort, if we were to endure seven years, or even three and half years, of tribulation it would not bring comfort but grief. Then in chapter 5 we read, "But you, brethren, are not in darkness, so that this Day should overtake you as a thief. You are all sons of light and sons of the day. We are not of the night nor of darkness." (verses 4,5) and again, "For God did not appoint us to wrath, but to obtain salvation through our Lord Jesus Christ," (verse 9). The tribulation is a time of darkness and we are not a part of that darkness and we are not a part of that wrath. In II Thessalonians and chapter 2 we read of how there is coming a lawless one, who is the antichrist. But before he can come there is a restrainer in the world that restrains evil and must be removed before this lawless one can have his way. The restrainer is the Holy Spirit who indwells all those who belong to the Lord Jesus. When the Holy Spirit departs from this earth He will take His Saints with Him because He is in each one of them. Once the Holy Spirit departs, the dam holding back evil will break and the tribulation will begin. So then after the rapture there will be seven years of tribulation and the end of this time of wrath will be when our Lord comes with His Saints and sets up His literal physical kingdom (Revelation 19:11-20:1-9).

With this understanding of the kingdom and dispensationalism it is time to give some thought to how the word *kingdom* is used in the New Testament Epistles. Up to this point we can see how the message of the kingdom is primarily to Israel and refers to a literal, physical kingdom that will be set up by the King that Israel rejected. At the same time we have seen how Lord also speaks of the kingdom as a mystery and something that is spiritual. Therefore, it is not surprising to find the word *kingdom* in the Epistles to the Church where this can create some confusion and lend credence to referring to the kingdom as synonymous with the Church.

In the book of Acts we see references to the kingdom as the Apostles preached the kingdom. For the most part this was in reference to the Jews where it was important to explain to them that Christ is the Messiah and the king promised by the prophets. The Jews continued to reject this message and the emphasis shifted to the Gentiles. The kingdom is still mentioned in the preaching to the Gentiles (Acts 14:22) such as we see at Ephesus (Acts 20:25). We are not told what this message is specifically. In Romans 14:17 we read, ". . . for the kingdom of God is not eating and drinking, but righteousness and peace and joy in the Holy Spirit." This would certainly indicate a spiritual kingdom is in view. Again in Colossians 1:13 we read "He has delivered us from the power of darkness and conveyed us into the kingdom of the Son of His love." This is what takes place at the time of our new birth we are taken from darkness into light or the kingdom of our Lord who is the light. However, overwhelmingly, the Epistles speak of the kingdom as that which is future. It is that ultimate goal that we go through various trials to enter eventually. The following is submitted as an example:

> "Do you not know that the unrighteous will not inherit the kingdom of God?" (I Corinthians 6:9). An inheritance is that which is future, the unrighteous do not have this inheritance.

". . . envy, murders, drunkenness, revelries, and the like; of which I tell you beforehand, just as I also told you in time past, that those who practice such things will not inherit the kingdom of God." (Galatians 5:21).

"For this you know, that no fornicator, unclean person, nor covetous man, who is an idolater, has any inheritance in the kingdom of Christ and God." (Ephesians 5.5)

". . . that you would walk worthy of God who calls you into His own kingdom and glory." (I Thessalonians 2:12). We are to order our conduct in a worthy manner today because of our great salvation and in view of what we shall be in the future.

". . . which is manifest evidence of the righteous judgment of God, that you may be counted worthy of the kingdom of God, for which you also suffer." (II Thessalonians 1:5). We will certainly suffer on our way to kingdom as we exercise the practice of the kingdom in the here and now.

"I charge you therefore before God and the Lord Jesus Christ, who will judge the living and the dead at His appearing and His kingdom." (II Timothy 4:1)

"And the Lord will deliver me from every evil work and preserve me for His heavenly kingdom. To Him be glory forever and ever. Amen!" (II Timothy 4:18)

"Therefore, since we are receiving a kingdom which cannot be shaken, let us have grace, by which we may serve God acceptably with reverence and godly fear. For our God is a consuming fire." (Hebrews 12:28-29).

"Listen, my beloved brethren: Has God not chosen the poor of this world to be rich in faith and heirs of the kingdom which He promised to those who love Him?" (James 2:5)

". . . for so an entrance will be supplied to you abundantly into the everlasting kingdom of our Lord and Savior Jesus Christ." (II Peter 1:11).

The sum of what we have so far is that there is a physical, literal kingdom yet to come, in the mean time there is a spiritual, mystical kingdom today. The basic definition of the kingdom is that of a king who has sovereign rule over His kingdom. In this sense our Lord rules as king today. All authority has been turned over to the Son ". . . as You have given Him authority over all flesh, that He should give eternal life to as many as You have given Him." (John 17:2). It is important to keep in mind that the most important aspect of our Lord's reign is that He gives eternal life and that He executes judgment (John 5:27). All through Scriptures we see the sovereignty of God at work. He raises up and takes down rulers and their kingdoms, ". . . till you know that the

Most High rules in the kingdom of men, and gives it to whomever He chooses." (Daniel 4:25). He provides the basic necessities for life upon both the just and the unjust, ". . . that you may be sons of your Father in heaven; for He makes His sun rise on the evil and on the good, and sends rain on the just and on the unjust." (Matthew 5:45). The powers that be are servants of God, "Let every soul be subject to the governing authorities. For there is no authority except from God, and the authorities that exist are appointed by God . . . . For he is God's minister to you for good." (Romans 13:1,4). Then one day every knee will bow and every tongue will confess that Jesus is Lord whether they want to or not (Philippians 2:9-11). Over all that is going on, God is sovereign and is moving all things along according to His will and to His glory.

Within the sphere of this spiritual kingdom the Church, that is all believers, are those who particularly bow down to His sovereign will. It is the members of the Church that order their conduct in a manner that is worthy of the reign of the King and the Lordship of Jesus Christ. At the same time God is allowing mankind to continue in rebellion against His rule i.e. to live in sin. They may seem to have their way today, they take advantage of the sunshine and rain but ultimately they will not inherit the kingdom. This might be pictured as a circle within a circle, the Church within the kingdom.

This is the reason why the Church will live out the principles of the kingdom such as found in the sermon on the mount (Matthew 5-8). This is also the reason why, even though the Olivet Discourse (Matthew 24-26) has to do with the tribulation period, we will still be watchful and will feed the hungry, give water to the thirsty, give shelter and clothing and visit those in prison. We do these things, not so that we can get into the kingdom because we are already in the Church and the kingdom is our inheritance in Christ, but because we are living out the kingdom in our lives. Therefore the prime objective of believers is to preach the gospel, everything that we do must point to the gospel and everyone who believes will have their place in the kingdom. The kingdom will not nor cannot be established through the works of men but only by the sovereign working of God. Let us focus then on the Church because that is who we are and what we are about.

CHAPTER ELEVEN

THE HOLY SPIRIT

As is true of most things in theology and the Bible there are various notions about the Holy Spirit and about His work. Inasmuch as the theme of this volume is evangelism, this chapter will deal primarily with the work of the Holy Spirit in evangelism. However, in order to do that there are certain facts concerning the Holy Spirit which need to be examined first if not at least in a cursory fashion. The word *Spirit* is translated from the word πνευμα (*pneuma*) which literally means wind. In John 3:6-8 we read, "That which is born of the flesh is flesh, and that which is born of the Spirit is spirit. Do not marvel that I said to you, You must be born again. The wind blows where it wishes, and you hear the sound of it, but cannot tell where it comes from and where it goes. So is everyone who is born of the Spirit." It is interesting that the same word translated spirit is also translated wind in these verses. This word is in the neuter gender but when attached to a pronoun, the pronoun is masculine.

The Greek word for spirit is *pnuema* (from which we drive English words that have to do with air, like "pneumatic" and "pneumonia") and is a neuter gender word. According to every normal rule of grammar, any pronound that would be substituted for this neuter noun would itself have to be neuter. However, in several places the biblical writers did not follow this normal procedure of grammar, and instead of using a neuter pronoun in place of the neuter noun *pneuma,* they deliberately contradicted the grammatical rule and used masculine pronouns. Indeed, they used three different kinds of pronouns, all in the masculine gender. This shows that they considered the Spirit to be a person and not merely a thing.[65]

It is important for us to understand that the Holy Spirit is a person and a person of the Godhead. The Spirit is not just an energy force in the cosmos but a person of the trinity who is of the same essence as the Father and Son. Charles Ryrie in his book on the Holy Spirit lists four reasons for the doctrine of personality. First, is that He has the **attributes of personality** which include **intellect** (I Cor. 2:10-11; Eph. 1:17), **emotions** (Eph. 4:30; Rom. 15:30), and **will** (I Cor. 12:11; Acts 16:6-11). Second, He performs the **actions of personality:**

The Spirit teaches—(John 14:26)

The Spirit testifies or witnesses—(John 15:26)

---

[65]   Charles C. Ryrie, *The Holy Spirit* (Chicago: Moody Press, 1965) 14.

The Spirit guides—(Romans 8:14)

The Spirit convinces—(John 16:7-8)

The Spirit restrains—(Genesis 6:3)

The Spirit commands and directs people—(Acts 8:29)

The Spirit performs miracles—(Acts 8:39)

The Spirit calls for special service—(Acts 13:2)

The Spirit sends forth into Christian service—(Acts 13:4)

The Spirit intercedes—(Romans 8:26)

Third, He receives the **ascriptions of personality**:

The Spirit can be obeyed—(Acts 10:19-21a)

The Spirit can be lied to—(Acts 5:3)

The Spirit can be resisted—(Acts 7:51)

The Spirit can be grieved—(Ephesians 4:30)

The Spirit can be reverenced—(Psalm 51:11)

The Spirit can be blasphemed—(Matthew 12:31)

The Spirit can be outraged—(Hebrews 10:29)

Fourth, He contradicts the **accidence of personality** (Accidence—"the rudiments of grammar") as seen in the quote above.

In John 16:13-14 the masculine demonstrative pronoun is used for *pneuma*.

In John 15:26 and Ephesians 1:14 the masculine relative pronoun is used for the neuter noun *pneuma*, Spirit.

In John 16:7-8 the masculine personal pronoun is used in place of the neuter noun for Spirit. [66]

Ryrie goes on to give four proofs with scriptures which demonstrate the deity of the Spirit. The first of these is **appellations**, "The fact that the Holy Spirit bears divine names is a proof of His deity."[67] The second one is that the Holy Spirit has **attributes:**

The Spirit is said to possess omniscience— (I Corinthians 2:11-12)

The Spirit is said to possess omnipresence— (Psalm 139:7)

The Spirit is said to possess omnipotence— (Job 33:4)

The Spirit is said to be truth—(I John 5:6b)

The Spirit is called the Holy Spirit—(Luke 11:13)

The Spirit is said to be a life-giver (Romans 8:2)

The Spirit is said to possess creative wisdom— (Isaiah 40:13)

The third one is **Actions:**

The act of creation (Genesis 1:2)

The act of inspiration (II Peter 1:21; II Timothy 3:16)

The act of begetting Christ (Luke 1:35)

The work of convincing—(John 16:8)

---

66   Ibid., 11-14.
67   Ibid., 17.

The work of regenerating—(John 3:6)

The work of comforting—(John 14:16)

The work of interceding—(Romans 8:26)

The work of sanctifying—(II Thessalonians 2:13)

The fourth one is **associations** which include His association with Jehovah (cf. Acts 28:25 and Isaiah 6:1-13; cf. Hebrews 10:15-17 and Jeremiah 31:31-34); His association with God (Matthew 12:31-32; Acts 5:3-4); and His association with the Father and the Son (Matthew 28:19; II Corinthians 13:14).[68]

Therefore we can state with confidence that the Holy Spirit is a person and that the Holy Spirit is God. Now we will examine the role of the Holy Spirit in our salvation. We know that our salvation begins with the Word of God, that faith comes by hearing and hearing comes by the Word of God (Romans 10:17). Our Bible, the Scriptures, are the inspired Word of God and this inspiration is by the Holy Spirit, "All Scripture is given by inspiration of God," (II Timothy 3:16), ". . . for prophecy never came by the will of man, but holy men of God spoke as they were moved by the Holy Spirit." (II Peter 1:21). It is clear that the writers of the Scriptures were allowed to insert their own personality and even include personal notes and greetings. But their thoughts and words were carried

---

68   Ibid., 17-20.

along by the Holy Spirit so that every word comes from God.

The Holy Spirit had a role in the very beginning of the salvation story as the Lord Jesus was conceived by the Holy Spirit, "And the angel answered and said to her, 'The Holy Spirit will come upon you, and the power of the Highest will overshadow you; therefore, also, that Holy One who is to be born will be called the Son of God." (Luke 1:35). The Holy Spirit led Him into the wilderness to be tempted by the Devil (Luke 4:1). Even though our Lord is God Himself, still He quoted from Isaiah 61:1 where He read, "The Spirit of the Lord is upon Me, Because He has anointed Me to preach the gospel to the poor;" (Luke 4:18). It is clear that the Son and the Holy Spirit worked together in perfect harmony doing the will of the Father.

The sacrifice of our Lord, the atonement for our sins was accomplished by the entire trinity, ". . . how much more shall the blood of Christ, who through the eternal Spirit offered Himself without spot to God," (Hebrews 9:14). The offering of the Son to the Father was through the Holy Spirit. Without the working of the Holy Spirit there is no salvation.

The preaching of the gospel and the power of the gospel is by the Holy Spirit. We see through the book of Acts how that the Holy Spirit directed the course of the ministry of the gospel. For example we see how that when Paul would have gone on into Asia they were forbidden by the Holy Spirit and ended up going to Macedonia instead (Acts 16:6-10). We also see how that when those who believed the gospel, they do so as they were empowered by the Holy Spirit, "And as many as had been appointed to eternal life believed." (Acts 13:48) and "The Lord opened her heart to heed the things spoken by Paul" (Acts 16:14).

We see the working of the Holy Spirit in regeneration particularly in John chapter 3. This text was examined in the previous chapter in relation to the kingdom, but now it is to be examined in relation to work of the Holy Spirit. Our Lord had just told Nicodemus that one must be born again before one can even see the kingdom of God. Nicodemus asked how can a person be born again when they are old. The answer our Lord gave is, "Most assuredly, I say to you, unless one is born of water and the Spirit, he cannot enter the kingdom of God. That which is born of the flesh is flesh, and that which is born of the Spirit is spirit. Do not marvel that I said to you, You must be born again." (John 3:5-7). Our Lord stated that the new birth was to take place by both the water and the Spirit. To speak of both water and the Spirit is not uncommon in the Scriptures. In Isaiah we read, "For I will pour water on him who is thirsty, And floods on the dry ground; I will pour My Spirit on your descendants,

And my blessing on your offspring;" (Isaiah 44:3) and then in Ezekiel, "Then I will sprinkle clean water on you, and you shall be clean; I will cleanse you from all your filthiness and from all your idols . . . . I will put My Spirit within you and cause you to walk in My statutes, and you will keep My judgments and do them." (Ezekiel 36: 25,27). These prophecies have to do with the coming kingdom but the point here is, that the pouring out of water and of the Spirit are spoken together. There is a need for cleansing and for the Spirit. Nicodemus was probably aware of the use of these terms.

In the New Testament there is an ongoing use of water and of washing in connection with the working of the Holy Spirit which will help us to see what is meant by it. In connection with the relationship between Christ and His Church, we read, ". . . that He might sanctify and cleanse her with the washing of water by the word." (Ephesians 5:26). It is important to note the relationship between the washing and the word. This washing and the work of the Holy Spirit in regeneration are brought together in Titus, ". . . not by works of righteousness which we have done, but according to His mercy He saved us, through the washing of regeneration and renewing of the Holy Spirit," (Titus 3:5). Our salvation is based upon the will of God and the Word of God, "Of His own will He brought us forth by the word of truth, that we might be a kind of firstfruits of His creatures." (James 1:18) and then again in I Peter, ". . . having been born again, not of corruptible seed but incorruptible, through the word of God which lives and abides forever," (I Peter 1:23).

A person must be born again. In our English Bibles we read, "Do not marvel that I said to you, You must be born again." This is the unfortunate part of our American English in that the word *you* can be either singular or plural. In the Greek text the first *you* is translated from σοι which is singular and the second *you* is translated from υμας which is plural. Nicodemus was not to marvel that everyone must be born again. This is both an elementary and a profound truth. Arthur W. Pink expressed the imperative necessity of the new birth as follows:

> The new birth is an imperative necessity because the natural man is altogether devoid of spiritual life. It is not that he is ignorant and needs instruction: it is not that he is feeble and needs invigorating: it is not that he is sickly and needs doctoring. His case is far, far worse. He is *dead* in trespasses and sins. This is no poetical figure of speech; it is a solemn reality, little as it is perceived by the majority of people. The sinner is spiritually lifeless and needs quickening. He is a spiritual corpse, and needs bringing from death unto life. He is a member of the old creation, which is under the curse of God, and unless he is made a new creation in Christ, he will lie under that curse to all eternity. What the natural man needs above

247

everything else is life, Divine life; and as birth is the gateway to life, he *must* be born again, and except he be born again, he *cannot* enter the kingdom of God. This is final.[69]

Concerning the new birth Pink goes on to state that, ". . . It is not the removal of anything from the sinner, nor the changing of anything within the sinner; instead, it is the communication of something to the sinner. The new birth is the impartation of the new nature."[70] This impartation of the new nature is accomplished by the Spirit and the Word. This coming together of the water and the Spirit and then the link of water to the active working of the Word of God is explained well by William Kelly in the following:

> Water, in the Scripture, is habitually employed as the figure of the word of God applied by the Spirit. It may be used also for the Spirit Himself in His own power; but still I need not point out the close connection there is between these two thoughts. However, here we have the Spirit distinguished from it, and this shows us at once the reason of the difference. The water is mentioned because God would draw attention to the character of what is applied, to what deals morally with the man. He might not at first be aware that what

---

[69]   Arthur W. Pink, *Exposition of the Gospel of John* (Grand Rapids: Zondervan, 1945) 113.

[70]   Ibid., 114-115.

made him sensible of his uncleanness was the Spirit of God. There must always indeed be in the soul, whenever the Holy Ghost thus acts, a consciousness that there is a dealing of some sort. In a word there never is or can be unconsciousness where there is a real operation of God. But then a man might in nowise comprehend that it is the Spirit of God; but this he knows full well, that the word judges him—that it brings him in as guilty and altogether unfit for the presence of God. Thus, "water" is the expression of the word dealing morally with the soul, convicting the man of being unclean, and not merely, cleansing. It is a question at first of the impartation of a new nature that the man had not before.[71]

Just as the Holy Spirit energized creation (Genesis 1:2) the Holy Spirit imparts this new life into all who believe upon the name of the Lord Jesus Christ. If this was all that the Holy Spirit accomplished, it would be enough but consistent with the grace of God, there is more.

The Gospel of John provides even more teaching on the Holy Spirit. In the upper room discourse our Lord gave comfort to His disciples and spoke to them of a helper, the Holy Spirit who would be sent to them by the Father (John 14:26). The word *helper* is translated from the word παρακλητον or as so often transliterated *paraclete*. It literally means to come along side (para) then

---

[71] William Kelly, *Lectures on the New Testament Doctrine of The Holy Spirit* (Denver : Wilson Foundation) 8-9.

to assist or act as an advocate. It is the one who comes to give assistance or to hold one up. Our Lord has ascended up into glory to be seated at the right hand of the Father but has sent the Holy Spirit to hold us up until we too ascend up to glory to be forever with the Lord. We also see that one of the ways in which the Holy Spirit helps us is to teach us. As we study the Scriptures we can count on the Holy Spirit to guide us into truth and to bring to remembrance the things that we have learned. The Holy Spirit has provided teachers for the Lord's people and uses these teachers to help us learn along with His internal guidance so that we can compare spiritual things with spiritual things to discover the truth.

Then as our Lord made His way to the Garden of Gethsemane through the streets of Jerusalem He continued to teach His disciples, including teaching on the Holy Spirit. He taught them that it was good that He was going away because the helper, the Holy Spirit will then come. In the world the Holy Spirit will do three things; ". . . convict the world of sin, and of righteousness, and of judgment; of sin, because they do not believe in Me; of righteousness, because I go to My Father and you see Me no more; of judgment, because the ruler of this world is judged." (John 16:7-11). The light of this glorious gospel has penetrated the darkness and it is convicting to the world. I am convinced that the number one reason why a person becomes an atheist is because they do not want to be confronted with their sin and in particular the sin of rejecting Christ. They want to be able to do what ever they want without accountability or restraint. Our Lord went on to elaborate more about the work of the Holy Spirit in our lives. He will guide us into all truth; He will not speak on His own authority but will speak that which is consistent with the whole trinity; He will tell of things to come and He will glorify our Lord Jesus (John 12-15). This passage is basically telling us that the Holy Spirit will inspire the Scriptures so that they are the Word of God. The Holy Spirit and the water coming together in the Word of God and the action that is then produced by the Spirit in the lives of the Saints to the glory of God. There is nothing here that brings glory to man, the believers are helped and guided along to understand and to act upon the Word of God all to the glory of God. This is not for self edification and to provide the Lord's people with a warm fuzzy.

In Ephesians chapters one and two we see the work of the Holy Spirit in our salvation. We see how the Holy Spirit is involved in our believing and then in our security, "In Him you also *trusted,* after you heard the word of truth, the gospel of your salvation; in whom also, having believed, you were sealed with the Holy Spirit of promise, who is the guarantee of our inheritance until the redemption of the purchased possession, to the praise of His glory." (Ephesians 1:13-14). Ephesians chapter one begins by stating what the trinity has done on our behalf; what the Father has done in verses 3 to 6, what the Son has done in verses 7 to 12 and then what the Holy Spirit has done in verses 13 to 14. When a person believes on the Lord Jesus they are sealed with the Holy Spirit. As already noted in a previous chapter this seal designates both ownership and security. The Holy Spirit is also the guarantee of our salvation. It is important to keep in mind that the way in which we know that we have the Holy Spirit and that this function has taken place is because the Word of God says so, it is the Spirit and the Water working together. There is no need for a supernatural or emotional experience, we know because the Bible says so. In chapter two of Ephesians we see the working of the Holy Spirit in verses 4 to 5, "but God, who is rich in mercy, because of His great love with which He loved us, even when we were dead in trespasses, made us alive together with Christ (by grace you have been saved),". The reason why we know that it is the Holy Spirit at work here because it is as we have already seen, the Holy Spirit is the one who energizes and gives life and is the one who makes the new birth possible. Therefore, with this understanding, it is the Holy Spirit that has made us alive and seated us in the heavenly places in Christ. This connects to the fact that the Holy Spirit is our guarantee, He is the one who has seated us in the heavenly places. As far as God is concerned we are already in heaven, it is a done deal.

When the Holy Spirit came to us He came bearing gifts. In Ephesians chapter 4 we read, "When He ascended on high, He led captivity captive, And gave gifts to men." (verse 8). Then in I Corinthians chapter 12 we have the teaching of spiritual gifts and of the one body of Christ. The text begins with, "Now concerning spiritual gifts . . . ." (Verse 1). There are many different types of gifts which are distributed to the body of Christ but there is one Lord and it is the manifestation of the Spirit given to each one to profit all the body. It is important to note that the gifts are given for the profit of all, they are not given so that the individual can be assured of their salvation or to feel good about themselves. The gifts are given for ministry to others. In order for there to be one body there must be the baptism of the Holy Spirit.

"For by one Spirit we were all baptized into one body . . . and have all been made to drink into one Spirit." (I Corinthians 12:13). Here it is important to carefully note the grammar, the words that are used. The phrase ". . . *by one Spirit* . . . " is translated from ". . . εν ενι πνευματι . . ." which is *in one Spirit*. Literally the phrase is that the Holy Spirit is the medium in which the baptism occurs rather than being the agent. Carson explains it this way:

> In the other six instances, related to the prophecy of John the Baptist, Christ as the agent does the baptizing, and the Holy Spirit is the medium or sphere in which we are baptized. Moreover whenever the verb *baptize* is used in the New Testament, it is the medium of the baptism—water, fire, cloud, and so forth—that is expressed using this preposition εν (en), not the *agent*. [72]

The verse goes on to state ". . . *we were all baptized into one body* . . ." which is translated from ". . . εις εν σωμα εβαπτισθημεν . . ." which is *unto one body were baptized*". The preposition eis is most often translated *unto* rather than *into*. Then the verb baptized is an aorist indicative which is most often translated as a past tense. The word *baptized* is to dip or to submerge. Therefore, a more literal reading would be that we all were submerged in the Spirit unto or with a view toward being one body. This past act of submerging all the members in the Spirit which then becomes one body occurred at Pentecost when the Spirit came to earth. Each time a person places their trust in the Lord Jesus they appropriate the fact that they are submerged in the Spirit and are now a member of this one body of Christ. The Spirit is in us all and we all are

---

[72] D.A. Carson, *Showing The Spirit* (Grand Rapids: Baker Book House, 1987) 47.

in the Spirit. This is all a part of our great salvation which originates from the grace of God and is appropriated by faith and faith alone. There is no separate baptism of the Spirit accompanied by speaking in tongues or by the manifestation of any of the gifts.

The primary teaching of I Corinthians 12 concerns itself with the body of Christ. The members of this one body have been submerged into the Spirit with the result that each is now a part of this one body. Each of these members has a place and a function within the body. No function is more superior or more needful than any other function. "But now indeed there are many members, yet one body. And the eye cannot say to the hand, 'I have no need of you'; nor again the head to the feet, 'I have no need of you'." (I Corinthians 12:20-21). In order for each member to carry out their function each one has been gifted by the Holy Spirit. Most often the gift or gifts given by the Holy Spirit will match the personality, skills and ability of the member. There are times when the Holy Spirit will gift someone in an area in which they would ordinarily not function but now gifted and led by the Spirit they excel.

There are four passages of Scripture where the gifts of the Holy Spirit are listed; they are found in Romans 12, I Corinthians 12, Ephesians 4 and I Peter 4. The most extensive list is in I Corinthians 12. A complete list of these gifts would look something like the following:

List A	List B
Teaching	Apostle
Ministry	Prophet
Exhortation	Word of Knowledge
Giving	Word of Wisdom
Ruling	Gift of faith
Show mercy	Gift of healing
Govern	Gift of miracles
Evangelist	Gift of Discernment
Pastor	Gift of tongues
Helps	Interpretation of tongues

There are 20 gifts in all and you will notice that I have placed them into two columns labeled A and B. I am one of those who believe that there are present gifts and past gifts. The present gifts are those in list A and the past gifts in list B. There are others who do not believe in *past gifts* that all the gifts are present and active today. There are still others who believe that there are *past gifts* but do not believe in as many as shown in list B. The reason why there are *past gifts* is because there are certain gifts which were foundational in nature and were used to get the Church started before the Canon of Scriptures were completed. Just as our Lord did miracles to prove that He was the promised Messiah, the Apostles did miracles to establish the truth of their message that the Lord Jesus is the Messiah and that all who place their faith in Him will have everlasting life and will be a member of His body, the Church.

The gospel was first preached to the Jew and the Jew required a sign, "Then some of the scribes and Pharisees answered, saying, 'Teacher, we want to see a sign from you.' But He answered and said to them, 'An evil and adulterous generation seeks after a sign,'" (Matthew 12:38-39). Notice how it is an evil adulterous generation that seeks a sign, in stark contrast, the Church is a holy bride (Ephesians 5:25-27). This is the reason why we read, "For Jews request a sign . . ." (I Corinthians 1:22) and "Therefore tongues are for a sign, not to those who believe but to unbelievers;" (I Corinthians 14:22). It is not hard to see who the unbelievers are in this verse as we see that each time tongues were used as a sign it was when Jews were present; at Pentecost (Acts 2:4-6); to prove that Gentiles could also be saved (Acts 10:44-48); and finally that those who had been baptized with the baptism of John could now be saved by faith in Christ (Acts 19:1-7). It is interesting that speaking in tongues is referred to only three times in the book of Acts. Clearly there are those who give much more importance to speaking in tongues than does the inspired Word of God. All the other times in the book of Acts when conversions are spoken of, there is no speaking of tongues.

The Apostles and Prophets laid the foundation for the Church. The Apostles were more than those who were sent with a message, they were endowed with special authority, they had gifts to vindicate their authority and their message, and they had seen the risen Christ i.e. they lived during that special period of time. The words spoken by the Apostles were often quoted from the Old Testament prophets, also there were prophets to tell the Church which would later become the Scriptures. Upon the completion of the Scriptures we no longer need Apostles because their authority now comes from the inspired Scriptures. Prophets are no longer needed because all that we need for life and practice is found in the inspired Scriptures. The need for someone to teach us the Scriptures or to "forth tell" the Word is provided by gifted teachers. The *word of knowledge* and the *word of wisdom* are now found in the inspired Scriptures, for example "For this reason we also, since the day we heard it, do not cease to pray for you, and to ask that you may be filled with the knowledge of His will in all wisdom and spiritual understanding;" (Colossians 1:9). The gifts of faith, healing and miracles, discernment and tongues have all been replaced by the inspired Scriptures as the Scriptures are able to vindicate themselves without the help of these sign gifts. As to faith, the just shall live by faith (Romans 1:17; Galatians 3:11; Hebrews 11:38).

There is no need for a special endowment of faith as there is enough faith available for us all to do the ministry the Lord has given us to do.

Concerning *speaking in tongues* it is worthy of note that the word *tongues* is translated from the word γλωσσαις (*glossais*) which literally means language. The word translated *language* is from the word διαλεκτω (*dialekto*) which literally means dialect. In other words on the day of Pentecost those in attendance heard the gospel, not only in their own language but in their right dialect of that language. The only time in which a language is an unknown language is when the person hearing it does not know that language. If a person started praying in Spanish, they might be praying in the spirit but to me it would not be understood and would not be fruitful. The Apostle Paul pointed out to the Corinthians that he spoke more languages than they did and so if he wanted to keep them ignorant of what he was saying, he could use a language that was unknown to them. However, his purpose was to edify or build up the Saints, not to impress them with the languages that he knew. There are those who say that they pray with the language of angels which is why it sounds like babbling. Whenever angels spoke to anyone, it was always in the language the person understood. The reason for this is because language is for the purpose of communication. This is important for the edifying of others which is the whole point to the gifts of the Spirit. The first time that God gave men different languages was at the tower of Babel and was the result of sin. The second time was at Pentecost and it was so all might hear the gospel and then take that gospel to their own people groups.

It is important to keep in mind that just because there are gifts that are past does not mean that God is not doing miracles, healing, giving discernment, wisdom, knowledge or even giving a person a certain language in which they were not otherwise trained for a specific task. To make that conclusion is an absurdity based upon a false understanding of the sovereignty of God. Our God is more than capable of doing whatever He wants, whenever He wants to do it according to the good pleasure of His will. The point here is that these things are left to the prerogatives of God. For example, a man who was dying of cancer asked if I would lay hands on him and cure him? I asked him why he wanted to remain in this sin cursed earth filled with pain and sorrow rather than to go and be with the Lord? He thought for a moment and then asked if I could pray that he could spend one more Christmas with his family. He went home to be with the Lord the first week in January. It was not up to me whether the man lived or died, it was simply a matter of asking if he could live through this one last holiday, God in His sovereignty allowed it and then took him home. The gifts of the Spirit are not to be used as some magic wand that we whip out whenever it suits us.

This then goes to the point of when and for what reason do we use the gifts of the Spirit. The gifts in list A are, by definition, others oriented. One of several values of the local Church is that it is a place where we can come together and put on display the functioning of the body of Christ. Therefore, when we are together we can serve one another according to the gifts given to us by the Holy Spirit. This is as we read, "As each one has received a gift, minister it to one another, as good stewards of the manifold grace of God." (I Peter 4:10). This will also move out to when we are separated in going to one another individually, but it all starts when we are together. We also go out into our community and exercise our gifts as a blessing to everyone around us which then leads to conversation which then leads to the giving of the gospel.

In Ephesians chapter 4 we see another important aspect to the work of the Holy Spirit and to the gifts that He brings to the body. Remember that there are four places in which the gifts of the Spirit are listed i.e. Romans 12, I Corinthians 12, Ephesians 4 and I Peter 4. In Ephesians 4 we see the Son and the Holy Spirit working together to bring gifted men to the Church. In verses 7 and 8 we read, "But to each one of us grace was given according to the measure of Christ's gift. Therefore He says: 'When He ascended on high, He led captivity captive, and gave gifts to men.'" Our Lord is the conquering hero who leads captivity captive and gives gifts to men. The functional working of this is by the Holy Spirit. The gifts that are given in this list are apostles, prophets, evangelists, pastors and teachers. As argued earlier in this chapter there is no longer a need for apostles and prophets due to the completed Word of God. This leaves us with three very important and functional gifts. It is important to understand that these are not just gifts but rather are gifted men given to the Church. Evangelists are those who are particularly gifted in giving the gospel and leading people to Christ both locally and globally. When we think in terms of global evangelism we ought to be thinking in terms of global evangelists. The pastor is a shepherd or one who feeds and cares for the sheep. Inasmuch as feeding the sheep is done with the Scriptures, then being a teacher is closely connected to being a pastor. There are those who argue that these two are the same man because of an absence of the article in front of teachers. In other words the text states that there are *some* who are evangelists and some who are pastors but it does not say that there are some who are teachers. Therefore, we can very well think of these as being the same man, particularly in view of the fact that these pastors need to be teaching. At the same time we know that teaching is a gift in and of itself I Corinthians 12:28. There are those in the Church who are teachers, therefore a pastor will also be a teacher but a teacher may not always be a pastor. Just by way of observation it is easy to see how that there are pastors who have more of a pastor or shepherding ministry that is their priority. While there are others who have more of a teaching ministry that is their priority. There seems to be a stronger bent in either direction. This means that there needs to be other elders who can pick up the slack where there is a lack.

For the purpose of this volume the importance of this text is found in verse 12, "for the equipping of the saints for the work of the ministry, for the edifying of the body of Christ." These men not only evangelize, pastor and teach they train others to do what they do. The gifts of the Spirit are given with a view toward equipping and edifying the body of Christ. This truth is seen most particularly here in this text. These men should be involved in the training of the Saints to do the ministry to the extent that they are training up their replacement. The lack of commitment to this teaching is detrimental to the health and well being of the Church and to advancing the cause of Christ in our world.

It can be argued that this lack of commitment to this teaching hurts the Church in a number of ways which may not even be seen. But there are at least three areas in which it is most noticeable. The first is seen when a Church loses its pastor to retirement, death or they just decide to go on to bigger and better things. When this happens it kills the momentum of the Church, the teaching is done by outsiders and there is a loss of consistency, pastoral care suffers because now nobody knows who is supposed to pick up the slack, growth is stymied because people invariably identify with the pastor. The search for a new one becomes a search usually conducted in the same manner as any corporation would search for a new manager and can go on for a year or more. When one who fits the prescribed paradigm and has the right credentials is found, it will take at least another year for that person to get to know the people and the culture of the Church. If the Church would train their men for ministry, they could have someone ready to step in immediately and carry on the ministry given to them by the Lord.

The second area is in Church planting. The modern trend is to have a second campus or satellite Church. The preaching is sent over by way of a DVD or a satellite hook up. This is where the Church has decided that technology can be used because it is there, not because of any redeeming value to the ministry. Here it is not necessary to train up another pastor, the Church simply records the sermon of one and sends it over. Now if the pastor gets run over by a bus, there are two Churches that are adversely affected. If this technology is such a great idea then why not just have sermons on DVD or satellite for all Churches. Each denomination could pick out their best guy, record him (or her in some denominations) and send it out to all their churches. It is true that the Church of biblical times did not have this technology so there is the possibility that if they had it, they would have used it. On the other hand it might well be that this is the very reason why God ordained that the Son should come and the Church begins at this time in human history. The emphasis in the Bible is on people, not on the economics, technology or any other stuff but upon people. The gifts of the Spirit are given to people and people are to minister to people so that people will spread the good news to more people. No one person is irreplaceable in the Church, therefore it is our task to train others to do the work of the ministry so that the ministry might expand as result of this important doctrine.

The third area is where there are those who do not know when it is time to turn over their ministry to the one who has been trained to replace them. In keeping with this important doctrine of training others, we should be looking to see how we can train our replacement and have an exit plan. In sports it is often seen where an athlete is trying to compete when it is obvious to everyone around him/her that their body is no longer responding in the same way as it did when they were younger. The brain is telling the body to do things that the body is no longer capable of doing. In the ministry this is much harder to detect because people tend to think that it is honorable for a person to continue in ministry until they fall over dead. There are a number of problems here. Old people tend to be stuck in their ways and have good reasons for it. I readily admit that I prefer certain ways of doing things over what is going on in this culture. There is a time when older men need to step aside so that younger men with a new vision and greater vitality can step up and do the ministry. In terms of evangelism, appearances are important to reaching out to the lost. It ought to be clear that the world in which we live likes to see energy, enthusiasm, and creativity not some old fogey that is having trouble getting around physically or mentally. The presentation of the good news and what we have to offer as the Church should be presented by those who are at the top of their game rather than those who dwell at death's door. Even in the Old Testament provision was made for the priests who were to lay down their burden (Numbers 8:25). This does not mean that the aged do not have a place in the Church because they do. They are the keepers of the history and traditions of the Church. While the Church is moving forward into the future, it is good to remember the things of the past and to bring them together with the things of the present and future. They are behind the scenes imparting the wisdom gained from years of experience. Youth is always better served when they respect and take full advantage of the knowledge and wisdom of the aged. In order for this to work effectively the doctrine of training others in ministry needs to be taken seriously and implemented.

As we continue in the book of Ephesians we see how the text moves from our position in Christ in the heavenlies to our practical daily walk. The first three chapters of Ephesians have to do with our position and the last three have to do with our practice. As one would expect, the Holy Spirit continues to be involved. It has already been noted how the Holy Spirit has provided gifted men to advance the ministry by training others in chapter 4. In chapter 5 we see how we are to walk or order our conduct within the sphere of love, light, and in a straight line. The question could arise as to how does one do that? The answer is found in verse 18, "And do not be drunk with wine, in which is dissipation; but be filled with the Spirit," We are given a stark contrast between being drunk with wine and being filled with the Spirit. It is a matter of control, when one is drunk with wine, then one loses control to the wine. But when one is filled with the Spirit, one is controlled by the Spirit. This principle of being filled with the Spirit is also spoken of in terms of *walk* in the Spirit (Galatians 5:16,25) and being *led* by the Spirit (Galatians 5:18). The result is that we produce the fruit of the Spirit (Galatians 5:22-23).

The fruit of the Spirit is love, joy, peace, longsuffering, kindness, goodness, faithfulness, gentleness, self-control. Take note that this fruit has to do with living life, with character, and a godly example to everyone around us. It is not a grandiose display of slaying in the spirit or babbling incoherent nonsense. It is extremely important to understand that the manifestation of the Spirit is seen in the day to day lives of the Lord's people. It is in how we love one another and how we live in such a way that the world around us will glorify God and seek to know Him through faith in our Lord Jesus Christ.

Then there are things that we are not to do regarding the Holy Spirit. In Ephesians 4:30 we read, "And do not grieve the Holy Spirit . . . ." The Holy Spirit is a person and can be grieved with our lives when we order our lives in contrast to what is of the Spirit. As the Word of God and the Holy Spirit work together to bring about our regeneration, so also they work together to bring about our holy living. Again, in I Thessalonians 5:19 we read, "Do not quench the Spirit". We are not to hold the Spirit down or ignore His leading in our lives. This does not mean that we are more powerful than the Holy Spirit and that He is just going to sit in a little corner of our hearts until we give Him permission to have an influence in our lives. We must remember that He convicts of sin and He will convict us of our sin when we fail to yield to Him. From my own experience I have learned that it is not at all pleasant to grieve the Holy Spirit or to try and quench Him.

The Word and the Spirit are seen working together again in our battle against evil. In Ephesians 6 verses 11-17 we see that we are engaged in a spiritual warfare and that we have been provided with spiritual armor to withstand the assault of the evil one. The weapon that is included is the "sword of the Spirit, which is the word of God." We engage the enemy with the Word of God knowing that the Holy Spirit will use that to win the victory.

In this chapter it has already been noted that all Scripture is inspired by the Spirit. At the same time it is the Spirit that gives us understanding of the Word. In I Corinthians 2:13 we read, "These things we also speak, not in words which man's wisdom teaches but which the Holy Spirit teaches, comparing spiritual things with spiritual." The Holy Spirit is our guide in understanding the Word of God and the will of God and it is done in comparing spiritual things with spiritual things. I have also contended that the first book a student of the Scriptures should have is a good concordance. A concordance is a great help in finding passages of Scripture that are related to one another so that it easy to compare Scriptures with Scriptures to help us in our learning and understanding. This text goes on to state, "But the natural man does not receive the things of the Spirit of God, for they are foolishness to him; nor can he know them, because they are spiritually discerned." (Vs 14). Here we have two sides to the same coin. On the one side, the Holy Spirit uses the Word of God to convict of sin and to bring about faith in Christ. On the other side is that when we present spiritual things to the unbelievers, they do not get it. These are things that cannot be understood by the natural man and therefore it is all foolishness to them. It is always interesting to me to watch how Christians will go before a state government and pronounce with utter confidence that thus says the Lord. Those politicians in government do not have a clue as to what they are talking about. The same thing is true when Christians send off letters of protest to their representatives in the Federal government, they are writing spiritual things to men and women who have no idea what they mean because these are things that can only be discerned by those who have the Spirit of God within them. This does not mean that we do not continue to do these things but it does mean that we should not be surprised when the response is little more than duh.

There is a great deal more said about the Holy Spirit in the Scriptures but for the purpose of this volume we will now examine one last aspect of the Holy Spirit. In II Thessalonians 2:7 we read, "For the mystery of lawlessness is already at work; only He who now restrains will do so until He is taken out of the way." Here we read of one who restrains the lawless one from taking over. From what we know of the Holy Spirit through out all of Scripture, it is not difficult to see that this restrainer is the Holy Spirit. The Holy Spirit is one day going to leave this earth in accordance with the will and plan of God. When He leaves, He will take the Church with Him, this is what we call the rapture. During the seven years of tribulation, the Holy Spirit will work in much the same way as He did in the Old Testament. He will still have influence and will still lead people to Christ but will not seal up the believer as He did with the Church. Therefore, the last act of the Holy Spirit during this dispensation is to take the Church up into the clouds to meet the Lord in the air and we shall be forever with the Lord.

In 1900 an evangelist and faith healer by the name of Charles Parham began to teach that speaking in tongues is a sign of the baptism of the Spirit and faith healing a part of God's redemption. This set the stage for the Azusa Street Revival beginning April 9, 1906 in Los Angeles California. This revival was set in motion by an African American

Preacher by the name of William J Seymour a 34 year old son of a former slave. The revival included ongoing meetings consisting of speaking in tongues, healings, miracles and what is known as the sign gifts. Seymour had been a student of Parham, who criticized Seymour for his excesses. This revival is said to have lasted three years and is considered to be the catalyst for the Pentecostal and Charismatic movements. Their theology and practice may differ from church to church but they all share the belief in the supernatural gifts. This movement has the strongest appeal to people of color, immigrants and those of the poor and lower levels of the social economic class.[73]

The reason why this movement is important to the contents of this volume is because of the damage that has been done by this movement to the testimony of the Holy Spirit and to the gospel of our Lord Jesus Christ. The Holy Spirit has been made to look like an emotionally hysterical person who is a babbling idiot rather than being respected as God and the third person of the trinity. The movement is an embarrassment to the truth of the gospel and verges very closely to heresy particularly when one is told that they must speak in tongues before they can be sure of their salvation. Therefore, it is important that we maintain and teach a clear, biblical view of the Holy Spirit.

---

[73]   Azusa Street Revival, http://en.wikipedia.org/wiki/Azusa_Street_Revival#cite_ref-charismaticcentury_5-0

CHAPTER TWELVE

THE EMERGING CHURCH

This chapter is taken nearly word for word from my doctoral dissertation as noted below.[74] The reason why I have included this chapter in this volume is because we are living in a postmodern world and it is important to understand how we are going to structure our outreach in this culture. The emerging church is a movement which seeks to give an answer to the postmodern world. I will propose that the solution being offered is the wrong one.

The first problem with the emerging church movement is defining what it is, as it can be so many different things with so many twists and turns in its theology and dogma.

What can be known of the emergent church is that it is a Christian (Christian in the broadest sense of the word) movement which began in the late 20th century and the early part of the 21st century. This would mean that it is not a recent movement. The participants tend to cross

---

[74] Greg Koehn "The Value of the New Testament Church in a Postmodern World" (PhD diss., Louisiana Baptist University, 2013) 56-90

theological boundaries and may be described as Protestant, evangelical, conservative, liberal, Catholic, Anabaptist, Adventist, charismatic, or reformed. Therefore, it could be termed as an ecumenical movement. This movement can be found all around the globe but primarily in North America, Western Europe, Australia, New Zealand and Africa. The terms used in the movement help to define it and are significant in understanding the movement. Emergent church writers will speak of a conversation to emphasize that they are an ever developing dialogue. They speak of a praxis oriented life style that incorporates political and postmodern elements. They teach the need for being missional but that is not the same as what we usually think of as missions. It is an evolving life style that will invite others into this same life style with the result that a person will arrive at their own form of spirituality. The theme of postmodernism flows through out the writings of this movement as there is a definite correlation.

There is to be the deconstruction of worship, evangelism and Christian community and out of that will emerge this new and better thing. It is worthy of note that the word *deconstruction* is defined in Webster as:

> A method of literary criticism that assumes language refers only to itself rather than to an extratextual reality, that asserts multiple conflicting interpretations of a text, and that bases such

interpretations on the philosophical, political, or social implications of the use of language in the text rather than on the author's intention.[75]

In the case of the emerging church, the deconstruction is a move away from what has been the traditional norm for the Church to something that is more in keeping with the cultural norm at the time which would be postmodernism. Whenever someone promotes change it means that they believe change is necessary and therefore, there needs to be justification for that change. Unfortunately people come to a decision based upon their own limited experience and think that the whole Church is equally guilty. In this case there are several who have come to the opinion that the Church is in need of fixing and that the best way to fix it is to be more like the world. The emerging church is that which is emerging as the solution to the broken Church and that which is more in keeping with our postmodern times. The word *praxis* is used on a regular basis or *orthopraxis*. The meaning is simply that Christianity and the Bible need to be communicated in the field of experience. This is in keeping with "life style evangelism" and the "social gospel". The emphasis is on a person living out the gospel in their life and allowing the Holy Spirit to direct others

---

[75] *Merriam Webster's Deluxe Dictionary*, 1998, 10th ed., "deconstruction".

to Christ in their own unique way, based upon their experience within a spiritual paradigm.

A more clear definition of the emerging church is as follows:

> The emerging or emergent, church movement takes its name from the idea that as culture changes, a new church should emerge in response. In this case, it is a response by various church leaders to the current era of post-modernism.

> Although post-modernism began in the 1950s, the church didn't really seek to conform to its tenets until the 1990s. Post-modernism can be thought of as dissolution of "cold, hard fact" in favor of "warm, fuzzy subjectivity." The emerging / emergent church movement can be thought of the same way.

> The emerging / emergent church movement falls into line with basic post-modernist thinking— it is about experience over reason, subjectivity over objectivity, spirituality over religion, images over words, outward over inward, feelings over truth. These are reactions to modernism and are thought to be necessary in order to actively engage contemporary culture. This movement is still fairly new, though, so there is not yet a standard method of "doing" church amongst the groups choosing to take a post-modern mindset. In fact, the emerging church rejects any standard methodology for doing anything. Therefore, there is a huge range of how far groups take a

post-modernist approach to Christianity. Some groups go only a little way in order to impact their community for Christ, and remain biblically sound. Most groups, however, embrace post-modernist thinking, which eventually leads to a very liberal, loose translation of the Bible. This, in turn, lends to liberal doctrine and theology.[76]

The key thoughts to keep in mind here is that this movement is seen as a way to engage in contemporary culture which is postmodern. The result of postmodernism is the dissolution of cold hard facts otherwise known as relativism. In relativism there are no absolutes which begs the question, what do we do with the Bible? If there are no absolutes then the Bible, which is full of absolutes, will be marginalized so that the absolutes of the Bible are not really absolutes after all. It is interesting to see how this is done which is a discussion that will follow in the pages to come.

In an attempt to understand the emerging church it must be understood that the existing Church needs to be dismantled or at least the expression of the Church needs to be dismantled. It would be unfair to say that the movement wishes to dismantle that which the Lord said He would build in Matthew 16:18. The emerging church movement is focused upon how the Lord's people do

---

[76] Got Questions?org, "What is the emerging/emergent church movement?" http://www.gotquestions.org/Printer/emergent-church-emerging-PF.html [accessed on Oct. 30,2012].

church or the modern practice of church. The definition of church becomes the more universal and mystical definition of the Church. In order to dismantle the modern church it must be established that the church is wrong and is in need of being dismantled or reformed. Eddie Gibbs and Ryan Bolger seek to do this by pointing our certain salient points. Among these are as follows:

> The church must recognize that we are in the midst of a cultural revolution and that nineteenth-century (or older) forms of church do not communicate clearly to twenty-first-century cultures. A major transformation in the way the church understands culture must occur for the church to negotiate the changed ministry environment of the twenty-first century. The church is a modern institution in a postmodern world, a fact that is often widely overlooked. The church must embody the gospel within the culture of postmodernity for the Western church to survive the twenty-first century.[77]

The writers submit that from AD 313 until the mid twentieth century the Church occupied a central position in Western societies. However, at the mid point of the twentieth century the modern culture gave way to the post-modern and the Church lost its place of influence.

---

[77] Eddie Gibbs, Ryan K. Bolger, *Emerging Churches* (Grand Rapids: Baker Academic, 2005) 17.

The evidence that is cited are the studies that have been done to show that weekly church attendance in the U.K. is down to 8 percent while in the U.S. it is 40 percent and the majority of this is in the Bible belt or what is known as the red (Republican) states. In the blue states (Democrat) church attendance is more in line with the U.K. Therefore, "the church must 'de-absolutize' many of its sacred cows in order to communicate afresh the good news to a new world."[78]

The emerging church movement is calling for another reformation. The first reformation set up the Church to be analytical, disciplined and based upon the propositional truth of the Bible. Now there needs to be a second reformation which will, in many ways, be a return to the time before the first reformation with more of an emphasis on the senses.

This movement is based in both the post-modern culture and in the youth culture. It can be concluded that this movement began with the youth, what is known as Generation X or Gen Xers and the youth that have come after them. It is argued that the "Boomers are

the last generation that is happy with modern Churches."[79] It is interesting that the guitar is the instrument of choice. Beginning with the hippy

---

78   Ibid, 19.
79   Ibid, 21

279

movement in the 1960's to the present, "Americans love their guitars. They make the best ones, and they make them sound good. The sound of the Jesus revolution was guitar driven."[80] Actually the popularity of the guitar goes clear back to the days of the Wild West and the cowboys. Country music is guitar based. In the emerging church the guitar is still essential to the informal gathering of people who gather in a small room or house environment to gently sway to the chords of a guitar and experience the Holy Spirit (note: in this volume there is always the possibility of sarcasm).

It is possible to conclude that the need for the emerging church movement is found in the inability of the modern church to communicate effectively to the post-modern culture and is therefore in need of being dismantled. Whenever one dismantles one thing, particularly something like the Church which has been the center of Western cultural society for over 1600 years, then one would be wise to have something in its place. This then goes back to the question, "What is the Emerging Church?"

Gibbs and Bolger, along with their many resources, will be relied upon to give further definition to this movement which shall be referred to as "this movement". One of those considered to be a founder or at least the

---

[80]   Ibid, 26.

one who popularized the term "emerging church" with her website, is Karen Ward of the Church of the Apostles, Seattle. She wrote the following:

> The emerging church is being willing to take the red pill, going down the rabbit hole, and enjoying the ride. It is Dorothy not in Kansas anymore yet finding her way home. It is Superman braving kryptonite to embrace Krypton. It is sight seeking wider vision, relationships seeking expanded embrace, and spirituality seeking holistic practice. It is "road of destination" where Christ followers, formerly of divergent pasts, are meeting up in the missional present and moving together toward God's future.[81]

With such deep intellectual and theological insight as this, it is no wonder that this movement is so popular with the postmodern world. To be a part of this movement is really easy, you need only to wrap yourself up in mystical, mindless fantasy and all is well. At least all is well in your own mind; the next logical step would be to take a pill, red or otherwise (or needle, snuff, smoke or snort) to help you along in your mystical, "spiritual" experience.

Another writer of the movement, Kester Brewin, Vaux, London, expressed it in this manner:

> People talk of revival but fail to see that what needs reviving must be by definition dying—and

---

[81]   Ibid, 27.

we are serious about not wanting that to happen, not standing by to let it all just wilt. But we are also pragmatic. We cannot undertake a revolution—this is not God's style. We worship a God of evolving change. It will take time and generations and mistakes and strange beasts. But we will keep at it, not because we think we are somehow the "salvation" of the church—far from it—but because now we have tasted something of this reconfigured body and we simply cannot go back to pews and song sandwiches.[82]

There is some interesting terminology being used here that is worthy of note. First, there is the assumption that the Church is dying and God is not capable of keeping it alive in its present form, there must be a new form that emerges or else it will die. Second, the use of the word pragmatic is significant in that it is a very familiar word in the postmodern world. Pragmatic is whatever works or change is needed because what is being done is not working and therefore something that does work needs to be discovered. The analytical declaration of thus says the Lord no longer works so there is a need of something less analytical and not so propositional. Experience or praxis is now going to be the road to travel to the postmodern church.

---

[82] Ibid.

On the one hand the authors of this movement are firmly convinced that the modern ideas of church are not viable in this postmodern culture, but at the same time admit that the movement itself is diverse and fragmented which makes it difficult to explain or define.

Therefore there is a struggle to define the emerging church. The emerging church is not Gen-X, seeker, new paradigm, or purpose-driven churches. Even those forms are in need of radical reform because they do not communicate well to the postmodern world. The key to this movement is how it relates to the postmodern world, how it communicates and even how it blends in with present culture.

In the emergent church it can be argued that any unbeliever could easily line up with all these statements and be a part of the emerging church without ever actually trusting in the Lord Jesus for their salvation. It can also be noted that the movement is very ecumenical as Brian McLaren states, "In many ways, I have more in common with Catholics than I do with Protestants. Indeed, once one embraces many non-modern forms of Christianity, one may find that one feels more Catholic than Protestant."[83]

Gibbs and Bolger identify nine practices that define the emerging church. The first three of these are considered

---

[83]   Ibid. 38.

to be the core practices: (1) identify with the life of Jesus, (2) transform the secular realm, (3) live highly communal lives, (4) welcome the stranger, (5) serve with generosity, (6) participate as producers, (7) create as created beings, (8) lead as a body, and (9) take part in spiritual activities.[84] At first glance there does not seem to be much difference between what is emerging and the desire of most believers. However, upon closer inspection one can see difficulties.

Inasmuch as the first three of the nine practices of the emerging church are considered to be **core practices**, it will serve the purpose of this volume to think critically about these three practices. The first of these is to **identify with the life of Jesus**. This means to see the Lord Jesus as He lived within His culture and then take the model of His life and apply it to our postmodern life and culture. It needs to be noted that this author chooses to refer to the Lord Jesus while the authors of this movement, along with many others, choose to simply speak of Jesus. How does one identify with Jesus without ever confessing Him as Lord? To put it in the words of our Lord, "Ye call me Master and Lord: and ye say well; for so I am." (John 13:13).

The manner in which this movement seeks to identify with the Lord Jesus is in the way in which He lived, that He associated with the down and out, the disenfranchised

---

84   Ibid, 44-45.

and he healed the sick. The emphasis of this movement is on His life of miracles and healing rather than upon His preaching, His death and resurrection. Therefore it is all about the kingdom that has come and how the kingdom is available to all those who identify with the Lord Jesus. The focus is on the here and now rather than the life after death. "We have totally reprogrammed ourselves to recognize the good news as a *means* to an end—that the kingdom of God is here. We try to live into that reality and hope. We don't dismiss the cross; it is still a central part. But the good news is not that he died but that the kingdom has come."[85] The emphasis on the kingdom is so strong that there is no difference between the kingdom and the Church. The message is to live the kingdom rather than be a part of the Church.

At this point there is a need to compare this thinking with what the Bible actually teaches. When the Scriptures speak of identifying with Christ it is to identify with His death and resurrection. One word that is used to express this is the word *baptize* which is to submerge as we read in Romans 6, "Know ye not, that so many of us as were baptized into Jesus Christ were baptized into his death? Therefore we are buried with him by baptism into death: that like as Christ was raised up from the dead by the glory of the Father, even so we also should walk in

---

85 Ibid, 54.

newness of life. For if we have been planted together in the likeness of his death, we shall be also in the likeness of his resurrection:' (verses 3-5). The Bible is very clear that our identification with the Lord Jesus is in His death and resurrection, the cross is not only central but it is essential in how we live today as well as our life after death. The Christian life as promoted by this movement certainly appears to be one of a bunch of sweet hearts going around doing wonderful things for people, even a miracle or two, and then the ungodly will simply see how lovely it is to be a follower of Jesus Christ and they will become a part of this flower garden of a kingdom. It almost makes one want to paint flowers on a VW bug.

In stark contrast the Scriptures teach of a life of sacrifice, humiliation and death. Our Lord told His disciples, "And he that taketh not his cross, and followeth after me, is not worthy of me." (Matthew 10:38). Crucifixion is a bloody, painful and humiliating way to die and yet the cross is the sign of discipleship. The preaching of the cross has always been and always will be foolishness to those who are perishing (I Corinthians 1:18). The Apostle Paul (by inspiration of the Holy Spirit) saw his identity with the Lord Jesus as being identified with His sufferings, death and resurrection, "That I may know him, and the power of his resurrection, and the fellowship of his sufferings, being made conformable unto his death;" (Philippians

3:10). If there is a problem with the modern day Church it is in the manner in which it has forgotten to take up the cross and to declare thus sayest the Lord.

The emphasis placed on the kingdom in contrast to the Church must also be disputed. It is the Church that is God's ordained institution. The kingdom was offered to Israel (particularly in the book of Matthew) and this kingdom along with the King was rejected. It is the Church where both Jew and Gentile are brought together in one body in Christ. Our Lord is head over the Church (Ephesians 1:22) and it is in the Church where He is glorified (Ephesians 3:21). It can be seen that this is a point where this movement is not anything new but just amillennialism an/or postmillennialism revisited. This is where there is no literal millennial kingdom but what we have now is as good as it gets or we are going to create a kingdom for the Lord to come and take over as King. These views have been showed to be lacking in many areas, just as the emergent church is lacking.

The second of these three core practices is to transform the secular realm or **transforming secular spaces**. This involves the tearing down of the sacred/secular divide. "Emerging churches tear down the church practices that foster a secular mind-set, namely that there are secular spaces, times, or activities.

This begs the question, does this mean that we are to understand that these dualisms do not exist? Take for example the dualism of "the natural versus the supernatural", in I Corinthians we read, "These things we also speak, not in words which man's wisdom teaches but which the Holy Spirit teaches, comparing spiritual things with spiritual. But the natural man does not receive the things of the Spirit of God, for they are foolishness to him; nor can he know *them*, because they are spiritually discerned." (2:13-14). This is certainly not the only reference where the Scriptures speak to a difference between spiritual and natural or the spirit and the flesh etc. There are a number of dualisms that exist whether we like it or not. The fact is that many of the Lord's people go to a job that is secular because not everyone can be hired by the Church and that is the word we use to describe that employment. These same people will do volunteer work at the Church and that will be different from their secular employment at many different levels. Reading the Bible is different from reading a newspaper. At the same time one would be hard pressed to fine an evangelical who would disagree with the fact that our whole life is to be spent as unto the Lord and that we are spiritual people regardless of whether we are at work or at Church. This is a problem for people who live their lives in academia which is why Gibbs and Bolger cite Ockham, Duns, and Descartes.

Most everyday Bible believing evangelicals would not even know who these men are, let alone take anything that they said seriously. Nevertheless, to understand this movement it is important to understand this concern. More importantly, it is important to understand that while there is agreement in principle, the practice being presented by the movement does create problems.

There are specifically three areas where the movement is seeking to dismantle dualism and they are the **Bible, worship and evangelism**. What a person does with the Bible is key to understanding any movement or argument. This movement seeks to move from the systematic to the nonlinear. In modern theology the Bible is viewed as a singular book with a singular message that is developed from beginning to end. The modern Church saw the Bible as authored by God with answers to life's questions and problems, it was like a giant jig-saw puzzle where when you put all the pieces together you would have the complete picture. The problem is that some of the pieces are missing and some do not seem to fit. Therefore, truth became subjective and was rendered according to the person in power.

The view of these authors is slanted toward the perspective of a strict denominational hierarchy where truth is imposed by those in power according to the dictates of the denominational power structure. As noted in a previous chapter, most "evangelical churches" follow a democracy type of rule where it is near impossible to impose anything but rather everything is subject to a vote. Therefore, the goal of reducing absolute truth down to a collection of opinions of the ignorant has already been accomplished through the rule of the people.

Nevertheless, this movement in particular, is an assault on the historical view of the Bible and of propositional truth. The emphasis is placed upon the Bible as a collection of stories that make up one big story with all the stories to be viewed within their cultural context.

It is interesting how the evangelical way of looking at the Bible has endured for hundreds of years and how propositional truth has led millions to the Lord and has exerted a positive influence on the growth and posterity of the United States in particular and other countries of the world by extension is now considered to be a "restricted vision". This new unrestricted view with many interpretations is seen as freeing and true enlightenment. There is an old maxim among evangelicals that there is but one interpretation but many applications. This has worked well because people understood that there was a firm foundation of propositional truth set down upon which we could build our lives and our coming together as a Church. With this movement that foundation is being eroded.

Scot McKnight is a self proclaimed proponate and author in the emergent church movement. He wrote a book by the name *the Blue Parakeet* where he taught the view of the Bible as a story conducted within culture or many stories conducted within many cultures that somehow end up being one story that is to be translated within ones current culture. He begins by demonstrating how there is a great deal in the law of Moses that we do not keep today and would rather not keep even if we felt that we should. A good example of this is found in his list of "commands we mostly don't keep" taken from Leviticus 19:

1. Be holy because I, the Lord your God, am holy (19:2)

2. You must observe my Sabbaths. I am the Lord your God (19:3)

3. When you reap the harvest of your land, do not reap to the very edges of your filed or gather the gleanings of your harvest. Do not go over your vineyard a second time or pick up the grapes that have fallen. Leave them for the poor and the foreigner. I am the Lord your God (19:9-10)

4. Do not go about spreading slander among your people (19:16)

5. Do not plant your field with two kinds of seed. Do not wear clothing woven of two kinds of materiel (19:19).

6. Do not eat any meat with the blood still in it (19:26)

7. Do not cut the hair at the sides of your head or clip off the edges of your beard (19:27)

8. Do not . . . put tattoo marks on yourselves. I am the Lord (19:28).

9. Stand up in the presence of the aged (19:32)

10. Keep all my decrees and all my laws and follow them. I am the Lord (19:37).[86]

The point that he seeks to make and belabors for a number of chapters is that we do not do most of these commands if not all of them because we have rationalized our way out of keeping them. In his words, we pick and choose what we want to do or not do. He does not seem to be aware of passages of Scripture that distinguish between law and grace, the old covenant and the new covenant and the mystery of the Church found in the New Testament. For the purposes of this volume, only a cursory review will be attempted as this is a subject that could very well make up a whole other volume.

---

[86] Scot McKnight, *the Blue Parakeet* (Grand Rapids: Zondervan, 2008) 115-116.

For this cursory review, a few gleanings from the book of Galatians will suffice. This epistle written to the Saints in the region of Galatia presents an excellent treatise on how there is a difference between law and grace or between then and now. An excellent summary of this argument is found in verse 16 of chapter 2, "Knowing that a man is not justified by the works of the law, but by the faith of Jesus Christ, even we have believed in Jesus Christ, that we might be justified by the faith of Christ, and not by the works of the law; for by the works of the law shall no flesh be justified." The difference between the works of the law and faith in Christ is that justification can only come by way of faith in Christ. Therefore, there is a very important difference and contrast between law and faith. This difference is further explained as the argument proceeds through out the epistle. In chapter three, those who are under that law are cursed because they cannot keep the whole law and the law is used as a tutor to bring us to faith in Christ. Once a person sees that it is not possible to do all that the law demands they are forced to the conclusion that they must cast themselves upon the grace of God by faith. Chapter four begins with the analogy of child being no different than a slave in that they are both under law and in bondage. It is worthy of note of how the law is spoken of as bondage but faith is spoken of as liberation. To answer the question of why we do not keep the law today is very simple; we do not keep the law because it brings a curse and is bondage to us. To return to the law in any form, either that which is stated in the Old Testament or that which is created by Christians, is to return to bondage, "But now, after that ye have known God, or rather are known of God, how turn ye again to the weak and beggarly elements, whereunto ye desire again to be in bondage? Ye observe days, and months, and times, and years." (Verses 9-10). In contrast to keeping the law the injunction is to "Stand fast therefore in the liberty wherewith Christ hath made us free, and be not entangled again with the yoke of bondage." (5:1). Herein lies the tension of the Christian experience, how does one live in the liberty of grace while being holy?

McKnight moves on from the Old Testament to the New Testament beginning with the gospels and the words of Jesus. A couple of examples include the reciting of the Lord's Prayer and conversion into the kingdom. His argument concerning the Lord's Prayer is that grammatically the word *say* as in when you pray say, should be *recite*. Since many Christians do not see the need or have any desire to recite the Lord's Prayer, this means that we do not do what the Lord has commanded. In the first place there are many Christians who do recite the Lord's prayer and many more see this as an index prayer whereby we are given headings which we can then fill in beneath each one what is peculiar to our lives, time and place which would be in keeping with the blue parakeet argument. It should be worthy of note that the very disciples who were actually present when the Lord gave these instructions never pass on to the New Testament Church the value of reciting the Lord's prayer but they do say a great deal about praying. Again, the emphasis upon the kingdom over the Church is seen when McKnight states:

> As a second example, here is what Jesus requires or expects of those who want to enter the kingdom of heaven, all from the gospel of Matthew.
> They must have surpassing righteousness (Matthew 5:20)
> They must do God's will (7:21)
> They must become as a child in humility (18:3)
> They must cut themselves off from whatever is in the way (18:8-9)
> They must abandon riches (19:23-24)
> They must separate from the scribes and Pharisees (23:13)
> How do we "apply" these so-called entrance requirements of Jesus? To anticipate, we mostly don't apply them. We have discerned what they meant and how to make use of them in our world.[87]

---

[87]   Ibid, 125.

Note the charge made by McKnight that "we mostly don't apply them." How does he know that? Surpassing the righteousness of the Scribes and Pharisees is not that difficult.

Who made McKnight a prince and a judge? How does one apply these "so called" entrance requirements without applying them in principle? Since when are there entrance requirements to the Church? Once again, the men who were actually in attendance went on to write (by inspiration of the Holy Spirit) that justification will take place by faith in the Lord Jesus Christ. This person then receives the Holy Spirit who makes them a new creation in Christ Jesus who will then do good works as a result of the Holy Spirit's work in their lives. This righteousness far exceeds that of the Scribes and Pharisees who sought a righteousness that is of the law and not of faith. This is a critical error in this emphasis of the kingdom over the Church in that it places the emphasis on the doing of good works rather than on believing in Christ. Whereas the pattern of the New Testament is to first believe in Christ and then as a result do good works, "For by grace are ye saved through faith; and that not of yourselves: it is the gift of God: Not of works, lest any man should boast. For we are his workmanship, created in Christ Jesus unto good works, which God hath before ordained that we should walk in them." (Ephesians 2:8-10). In order for any person to understand or be able to in any way keep these sayings of our Lord, they will need to keep reading in the Epistles.

McKnight does proceed into the Epistles continuing his theme of how we tend to pick and choose what we want to obey in the Scriptures. He begins with a chapter on *discernment*. Since the Bible cannot be seen as statements of absolutes or propositional truth it is left to the readers and Church leaders to discern the meaning of the text in view of our present culture.

Note how the Bible is not seen as propositional, absolute truth here but as a book of guidelines to be discerned by both the individual and by the collection of individuals in the local church and the denomination. This is a key to understanding how the emergent church views the Bible and how it can be used to conform to the culture, specifically the postmodern culture. In other words if the Bible cannot change the culture then let the culture change the Bible.

Continuing on in his chapter on discernment, McKnight gives seven specific examples of where discernment was used within the biblical context. It is important to cite these and make some comment as they are an excellent example of how McKnight and the emerging church views the Bible and how one views the Bible is the most important component of any movement.

The seven examples are (1) Divorce and remarriage (2) Circumcision (3) The style of Christian Women (4) Sun-Centered or Earth Centered Cosmology (5) The death penalty (6) Tongues (7) All things to all.[88]

Anyone who has ever studied the Bible at all is familiar with the arguments concerning divorce and remarriage and the exception clauses. On the one hand our Lord teaches that there is to be no divorce as in Mark 10:11-12 but then on the other hand our Lord discerns an exception in the event of *sexual immorality* in Matthew 5:32. The apostle Paul also exercises discernment when he provides the exception of desertion in I Corinthians 7:15. Now, according to McKnight and others, the Church may discern that there are more exceptions such as abuse or immaturity. The Church now has the same authority as our Lord Jesus and as the apostle Paul writing the inspired text. What if the local Church is carnal, can it be depended upon to make good decisions on such questions? This continual move toward exceptions to the permanency of marriage, is this really the best direction for the Church to move? Once the Church begins to move away from what the Bible states it invariably creates more problems than it solves.

Circumcision is used as an example of the Church and Paul in particular exercising discernment that

---

88   Ibid, 131-144.

circumcision was no longer to be followed because in Christ the circumcision of the heart is more important than the outward circumcision of the flesh. This was not as a result of culture not wanting to be circumcised, it was the leading of the Holy Spirit in the writing of Scripture to make an important point between law and grace and between Israel and the Church. The Church is not in a position to add to Scriptures by dictating a new theological position. For example, most male children are circumcised for health reasons in our culture, do we now make it a spiritual principle as well? The answer is, of course not. The inspired Word of God has given us the timeless teaching on the subject that is good for every culture everywhere. Therefore, there is no need to bring circumcision into the Church because culture has decided that it is a good health practice.

The third of these has to do with the style of Christian women and in particular the text of I Peter 3:1-6.

This text contains three basic commands to women in first-century Asia Minor who had unbelieving husbands. They should:

- Submit to their non-Christian husbands in order to convert them.
- Avoid elaborate hairstyles and gold jewelry and fine clothing.
- Address their husbands with the word "lord".

In our day Christians do not insist that women refrain from nice clothing, expensive jewelry and do not insist that women call their husbands "lord". There are some groups of Christian who do follow these instructions as being literal from the Scriptures and one has to wonder what if modern day Christians did insist on these things? Just because Sarah called Abraham "lord" does not mean that all wives have to, there are certainly no men running around today who measure up to Abraham but some kind of respect certainly would not hurt either. But to answer the question of what would happen if Christians insisted upon such behavior, obviously this would make the gospel disagreeable and women in the postmodern age would rather go to hell than give up their expensive cloths and jewelry, much less call their husbands "lord".

The fourth example of "Sun-centered or Earth-centered cosmology" is even more of a stretch than the others. The contention is that the Bible teaches that the earth is the center of the universe and that it is flat. In our modern times we know that the sun is the center of our universe and that the earth is round. We are just so much smarter than the writers of the Bible. It is true that the Bible speaks of the four corners of the earth and that the earth is the center of attention for the Scriptures. However, the Bible also teaches that the earth is round such as in Isaiah 40:22, "It is he that sitteth upon the circle of the earth,". It is logical that the Scriptures would focus on earth because that is the focus of God in all of His creation. It is where the apex of creation, humans made in the likeness of God, live. To speak of the four corners is not a scientific treatise but a euphemism for teaching doctrine. This is one of many problems one has when one begins to view the Bible as an historical text book rather than the inspired Word of God.

The next two examples have to do with the death penalty and speaking in tongues. Both of these have Scriptures that can be used to defend the various positions concerning each. It is a question of whether or not to do it or not to do it i.e. to execute or not and to speak in tongues or not. The point of this volume is not the debate over what the Scriptures state which has been ongoing since the ink first dried but rather the problem of when a changing culture is used to interpret Scripture or is used to just ignore Scripture.

The seventh of these examples is Paul's statement that he has become all things to all men that he might reach some. The idea here is that Paul was willing to adapt to the culture while maintaining the truth of the message of the gospel. Most everyone will agree on this point, even the modern day Church. The point of contention is does the truth of the message change culture or does culture and the adapting to it change the truth of the message?

The fact is that the gospel that Paul preached changed the culture of the Roman-Greek world. It raised the esteem and place of women, it gave value to all human life regardless of their station in society and it brought people together from every race, kindred and language. This is the power of the gospel to change lives from the inside out and those changed lives will change their culture for the good of all. Paul became all things to all men but he never lost or diminished his boldness in preaching the gospel whether it was offensive or not. In fact he would write, by inspiration of the Spirit, that it is offensive (I Corinthians 1:20-31). The gospel has always been seen as foolishness and those who preach it as fools; this is not going to change by being more accommodating to the culture, assuming that the message remains the same.

McKnight ends his book with a treatise on the place of women in the Church. He argues that educated and spiritually gifted women should have a place in the preaching and teaching ministry of the Church. Once again, it is a matter of interpretation within the cultural context rather than a view of Scripture as absolute truth for all time in all places. This is how the emerging church views Scripture, a story that is still unfolding and being adapted by culture. The line between whether the culture defines the Bible or the Bible defines culture has been blurred so that it is nearly impossible to tell which it might be. Extreme caution needs to be enforced whenever there is a movement away from the Bible as being absolute truth. The blue parakeet may have just choked on its cracker.

The **second area of dualism** that needs to be broken down, according to this movement, is that of **worship**. The dualism here is that of the *secular* and the *sacred*. Music has become one of the most controversial of the various aspects of worship. Music has also become the focal point of worship to the extent that when it is time to sing and the instruments start to play it all begins with an announcement that it is time to stand and worship. The argument made by this movement is that the reformation was born in a literary age and therefore the focus was upon the written word rather than upon visual and sensual. It is argued that the reformation probably would not have happened if it were not for the printing press. On the other hand it could be argued that there would have been little use for the printing press if it were not for the printing of Bibles, commentaries and papers written on theological subjects. There is a great deal of evidence that the English language and reading was advanced by the printing of the King James Bible. People learned to read so that they could read the Bible. It was in the reading of the Bible that the gospel was spread and the reformation became successful in defeating the abuses of the Church of Rome. Thousands of languages have had their spoken language put into a written form so that they might read the Bible, thereby spreading education along with the Bible to the four corners of the earth (four corners being a euphemism not to say that the world is flat). The reformation came about by the grace of God because the Church had been ill treated by Rome and the truth of the gospel had been terribly compromised. To suggest a move back to the period prior to the Reformation is at the very least dangerous.

In the breaking down of the dualism of worship, this movement sees no difference between the *sacred* and the *secular* because all things have been made sacred. This is a movement to include all that is a part of the postmodern culture including the "visual and aural aspects of culture, including radio, TV, motion pictures, and the computer."[89] However worship in the emerging church is more than just the use of modern technology it includes paintings, slides, drawings, candles, tv, videos, anything that will enhance the visual and emotional experience.

The type of worship described here is like a giant warm fuzzy, a folding together of the physical earth and the Spiritual, the culture and the Church, no longer two but one in sweet harmony.

---

[89] Eddie Gibbs, Ryan K. Bolger, *Emerging Churches* (Grand Rapids: Baker Academic, 2005) 70-71.

The ongoing problem with this movement is that it contains elements of truth which cannot be argued and elements of compromise where there is no strict biblical basis for saying that this is not true or that it is sin. It is like the movement is dipping its toe into the waters of compromise and sin without plunging in headlong. Or it might be viewed as a kettle of water that is gradually heating up until the frog is too late in realizing that it has been boiled. There is no place in the Bible that states that modern technology cannot be used in our worship. It would be difficult to argue against the use of videos, power point, computers, etc. in worship including the teaching of the Word. God gave mankind the ability to invent these things and use them to His glory. Therefore it is not the use of the stuff that is the problem, but rather the problem must run deeper into the spiritual realm. Including icons, candles and artwork as a part of worship is interesting in that it was and is so much a part of the worship of the Roman Church. That style of worship was key for a people who were uneducated and illiterate. The success of the reformation was based upon the ability to read as much as it was on the printing press. Therefore, we might conclude from this type of worship that is based on the sensuous and upon ignorance? It certainly made some contribution to the sin of the church that demanded the reformation. Are we sure that we want to go back to pre-reformation days where we compromise with Baal and all the pantheism that went with it? If New Age practices and ecology become a part of our worship, how can we be sure of what is being worshipped, the creator or the creation?

There are at least three problems that need to be addressed concerning this version of worship. First, there is a dualism between the creator and creation which must be maintained. For example we read in Isaiah, "Those who make an image, all of them are useless, and their precious things shall not profit; they are their own witnesses; They neither see nor know, that they may be ashamed." (44:9). The text goes on to explain how foolish it is for a man to cut down a tree and out of part of it, he burns to cook his food and then from another part he carves out an idol and bows down to worship it. "And no one considers in his heart, nor is there knowledge nor understanding to say, 'I have burned half of it in the fire, yes, I have also baked bread on its coals; I have roasted meat and eaten it; And shall I make the rest of it an abomination? Shall I fall down before a block of wood?' He feeds on ashes; A deceived heart has turned him aside; and he cannot deliver his soul, nor say, 'Is there not a lie in my right hand?'" (vs 19-20). The Lord does not change His mind in the New Testament about such things,

"Professing to be wise, they became fools, and changed the glory of the incorruptible God into an image made like corruptible man-birds and four-footed animals and creeping things. Therefore God also gave them up to uncleanness, in the lusts of their hearts, to dishonor their bodies among themselves, who exchanged the truth of God for the lie, and worshiped and served the creature rather than the Creator, who is blessed forever. Amen." (Romans 1:22-25). The dualism between creator and creature must be maintained so that people will clearly understand who alone, is the object of our worship.

The second problem with bringing nature into the spiritual realm is that nature bears the curse. Regardless of the beauty of creation and how it points to a creator, it is not what it was when first created. Nature bears the marks of sin and the curse. The redemption of creation is linked to the redemption of man, "For we know that the whole creation groans and labors with birth pangs together until now. Not only that, but we also who have firstfruits of the Spirit, even we ourselves groan within ourselves, eagerly waiting for the adoption, the redemption of our body." (Romans 8:22-23). All the sacrifices of the Old Testament (note: Leviticus 23) point to the sacrifice of our Lord and this is the sacrifice that is pleasing to our God. "And walk in love, as Christ also has loved us and given Himself for us, an offering and a sacrifice to God for a sweet-smelling aroma." (Ephesians 5:2). He is the one who bore our sins in His own body on the tree and is the propitiation for our sins and He alone is to be the object of worship. If what we do does not point to Christ, then what we do is not worship.

The third problem comes from the rational that it is not the image that is being worshipped but the Lord or the spirit behind the image. This is a very fine line between worship of the Lord and panentheism. Pantheism is that God and the world are the same or a great oneness. "Panentheism is the belief that God is in the world the way a soul or mind is in a body;".[90] The Lord's people are to be taught proper theology so that they may reject this view. It is certainly not something to be encouraged as a part of our worship so that the ungodly might feel more at home.

The **third area of dualism** that needs to be broken down, according to this movement, is that of **evangelism.** Again, the dualism of the *sacred* and the *secular* is in view and especially that evangelism is not an event but an every day way of life.

This is basically another form of what is known as *lifestyle* evangelism. Basically, this is where Christians live out their "Christian" life and when people see this they inquire as to what that Christian has going for them,

---

[90]   Norman L. Geisler, *Christian Apologetics* (Grand Rapids: Baker Publishing, 1976) 193.

because they would like to have that as well. This is easy does it, low key and non-confrontational. Again, there are a number of problems with this view. In the first place no one will ever go to heaven by observing that Christians are a bunch of sweet hearts. In I Corinthians 15:1 we read, "Moreover, brethren, I declare to you the gospel which I preached to you, which also you received and in which you stand;" The gospel is declared and preached, in other words it must be verbalized. Some where in the course of making friends and influencing people the propositional truth of the gospel must be stated. Unless a person hears the Word, understands the Word and believes the Word they will not have everlasting life. Secondly, how often does someone come up and ask about being a Christian? Apparently in the emerging church it happens all the time which would explain why they meet in such small groups in houses. Ordinarily it does not happen and one reason is because Christians do not live lives that demonstrate holiness. In the ongoing effort to blend in and be a part of culture and not be offensive, a person cannot tell the difference between a believer and a non-believer. The Christian must know how to turn a conversation toward spiritual things and then move into giving the gospel. It is possible to be on the offense without being offensive with proper training. Thirdly, both our Lord and the Apostle Paul were confrontational.

When our Lord speaks of discipleship He speaks in terms of division and of a cross such as Matthew 10:34-39:

> Do not think that I came to bring peace on earth, I did not come to bring peace but a sword. For I have come to set a man against his father, a daughter against her mother, and a daughter-in-law against her mother-in-law; and a man's enemies will be those of his own household. He who loves father or mother more than Me is not worthy of Me. And he who loves son or daughter more than Me is not worthy of Me. And he who does not take his cross and follow after Me is not worthy of Me. He who finds his life will lose it, and he who loses his life for My sake will find it . . .

Our Lord speaks of coming with a sword and the Word of God is spoken of as a sword (Hebrews 4:12). A sword will cut, slash and divide. There will be many who are offended at the Word and will pursue their godless ways. There were those who departed from the Lord Himself and would not follow Him (John 6:66). There were others where our Lord did not commit Himself to them because He knew what was in their hearts (John 2:24). Our Lord speaks of a broad road that leads to destruction and a narrow road that leads to life (Matthew 7:13). There are few on the narrow road and many on the broad road.

The error of this movement is the assumption that people are naturally attracted to the gospel and if it is presented in a way which fits in with their cultural world view, then they will come to Christ. The Scriptures teach just the opposite that people are dead in trespasses and sins (Ephesians 2:1). The question that needs to be answered is, how many decisions does a dead man make? In Ephesians 2:2-3 we have what it means to be dead clearly defined. It means to order ones conduct according to the world, the devil and the lusts of the flesh. This doctrine of the total depravity of man is clearly outlined for us in Romans 1:18 to 3:20. This can be summarized with the words of Romans 3:10-11, "As it is written: "There is none righteous, no, not one; There is none who understands; there is none who seeks after God." Unless the Holy Spirit quickens or makes alive this dead person (Ephesians 2:4-5), they will not seek after God regardless of how seeker friendly a Church might be. Salvation is of the Lord.

People love to talk about how the Apostle Paul was all things to all men but they ignore how he was also confrontational in his preaching. At the very beginning of his ministry in Acts 13 while on the island of Paphos he was met by a sorcerer, a false prophet, Paul did not sit down with him, sip some tea and have an interactive discussion but rather said, "O full of all deceit and all fraud, you son of the devil, you enemy of all righteousness,

will you not cease perverting the straight ways of the
Lord?" (Acts 13:10). This was followed by his preaching
in the synagogue of Antioch where he outlined the history
of the people of Israel including the fact that they put
to death the one who was not deserving of death. He
then concluded with, "Therefore let it be known to you,
brethren, that through this Man is preached to you the
forgiveness of sins; and by Him everyone who believes
is justified from all things from which you could not
be justified by the law of Moses. Beware therefore, lest
what has been spoken in the prophets come upon you:."
(Acts 13:38-40). In Acts 14 Paul preached the gospel
in Lystra and was stoned and left for dead. Apparently
more than one person was offended. In Acts 17 the Jews
at Thessalonica caused a riot and accused Paul and other
believers of turning the world upside down. Rather than
blending in with culture, they clashed with culture and
turned the whole thing upside down. In Athens Paul went
into the market place, not to patronize with unbelievers
but to confront them with the Gospel. The result was
that he was placed front and center on Mars Hill where
he preached the Gospel concluding with, "Truly, these
times of ignorance God overlooked, but now commands
all men everywhere to repent, because He has appointed
a day on which He will judge the world in righteousness
by the Man whom He has ordained. He has given

assurance of this to all by raising Him from the dead." (Acts 17:30-31). Paul's message was not conciliatory but rather confrontational. He did not encourage men to find their own way of spirituality so as to enter the kingdom of God but declared emphatically the one way to the one God. The Apostle Paul did not blend into the crowd nor did he ever change his message to meet with the culture but in boldly declaring the truth of the Gospel he confronted culture and the culture of the Roman world was dramatically influenced. The best thing that we can do in the midst of this postmodern world is to do likewise.

The third of the **core values** is to **live highly communal lives.** Enough has already been written concerning the emerging church movement, nevertheless this chapter will conclude with more observation. Eddie Gibbs and Ryan Bolger in their book *Emerging Churches* call for the Church to be more involved in the community. Again, there is no disagreement there. Anyone with an elementary grasp of evangelism knows that the mission field of the Church is outside the boundaries of the building.

Again, there are certain elements here which are commendable and where believers can agree. It is true that the Lord Jesus Christ should be the center of everything that we do and that people ought to see Christ in us as we move about in the community. Our salvation does not stop with faith in Christ and wait for death or the rapture, we have been placed within the sphere of good works, "For we are His workmanship, created in Christ Jesus for good works, which God prepared beforehand that we should walk in them." (Ephesians 2:10). However, it is not the kingdom but the Church which is at center stage during this dispensation. It is important to emphasize again that the kingdom that our Lord preached was rejected and will be established, whether the Jews like it or not, at a future time after the tribulation. After our Lord's death and resurrection, Paul was given the task of making known the mystery of the Church that both Jew and Gentile along with all people would be united together, by faith in Christ, into one body, the Church. The Gospels are wonderful and have their place but it is in the Epistles where we learn of the Church and how we as believers should live. The Gospel is not this complicated web of kingdom requirements and kingdom lifestyle, the Gospel is believe on the Lord Jesus Christ and you will be saved (Acts 16:31).

CHAPTER THIRTEEN

A STRATEGY

In this chapter we want to take all that we have learned from the previous chapters and bring it together to form a strategy for Church growth and Church planting. This strategy is taken from my understanding of the New Testament and specifically the book of Acts. There are details to this strategy which are not found in the Scriptures but are based upon biblical principles and the common sense given to us by our Lord.

This strategy is built upon what I call the *Molecular Model*. There is a good chance that someone else has used that term in the past, I just do not know who it is, if I did know I would give credit where credit is due. The *Molecular Model* is based upon a growth strategy that expands to the point of overflowing the base Church. It is based upon a philosophy of Church planting rather than a mega-church philosophy. Each Church plant will be an autonomous Church but will have an intimate relationship with the base Church for purposes of training, encouragement and counsel. When the Church growth strategy explodes, the consequent Church plants will, by nature of their relationship to one another, form a fellowship of Churches for the purpose of mutual encouragement, evangelism (joint evangelistic events etc.), global evangelism (sending teams from the fellowship to plant Churches among unreached people groups) training (for every area of ministry including evangelism, teaching, discipleship, worship, shepherding etc.)

A Diagram of the Molecular Model may look something like this:

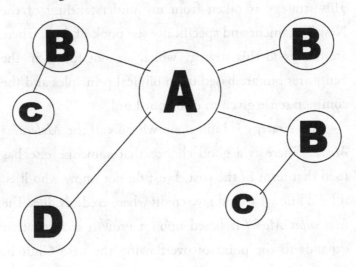

The base Church (A) plants Churches (B) they in turn plant Churches (C) then the base Church with the Church plants, plant a Church in an unreached people group (D). This will all happen quickly once the growth strategy explodes.

In Summary:

- Base Church explodes in growth
- Plants another Church
- Church plant explodes in growth and plants a Church
- Base church continues to explode in growth and plants another Church
- Church planting continues and together the Churches form a fellowship based upon their common theology, philosophy and growth
- Together the Churches plant a Church among an unreached people group

Theology:

1. The theological base for this strategy consists of the following 9 points:
2. That our Lord will build His Church Matthew 16:18
3. That the gospel has the power to change lives Romans 1:16; I Cor. 6:9-11
4. That we all are ambassadors for Christ to bring this life changing message Romans 10: 13-15; II Cor. 5:11-21

5. That salvation is of the Lord Jonah 2:9; Ephesians 1:3-14; e.g. Acts 13:48; 16:14

6. That the Church is God's ordained institution Colossians 1:18; I Timothy 3:15

7. That each member is gifted by the Holy Spirit to serve the body and advance the cause of Christ I Cor. 12:12-31; Ephesians 4:11,12

8. That it is possible for the body of Christ in its local form to be of one mind Ephesians 4:2; Philippians 1:27;2:2

9. That the Word of God is all we need for life and practice II Timothy 3:16,17

The local Church is established and built up by the Lord through the working of the Holy Spirit in the lives of the believers. Therefore, any strategy must be based upon this basic premise. Our responsibility as followers of Christ is to proclaim the message of Christ at every possible opportunity. Then to bring people into the safe harbor of the Church where they can be nourished and cared for in the Scriptures according to the gifts of the Holy Spirit. Philosophy and strategy must necessarily flow out from these 9 points of theology listed above.

Our Lord will build His Church through the working of His Spirit in the lives of His people who live by faith. When a person becomes a new creation in Christ they are a child by birth and a son by adoption which is one who is legally qualified and responsible to carry on the family business (Romans 8:14-17). The Church in the book of Acts understood this and went forward accordingly. First we see the mission of the Church in Acts 1:8 that the gospel is to move out in ever widening circle, hence the molecular model as diagramed above. Next we see the Church is large, 3,000 came to know the Lord on the very first day (Acts 2:4) and the Lord added to that number every day after (Acts 2:47). Another 5,000 were added on still another day (Acts 4:4). When we grasp the reality of the power of the gospel we will anticipate growth as the Lord adds to the Church. Since the Church is God's ordained institution, the focus is on the Church and its growth and ever expanding ministry. It is the Church that plants Churches which causes a chain reaction to where that Church plants a Church and that one etc. on and on to the uttermost parts of the earth. The Apostle Paul was sent out by the Church in Antioch (Acts 13:1-5) and it was to the Church that Paul reported (Acts 14:26-28). It is interesting to note that Paul's first missionary journey was from 44-46 AD; his second missionary journey was from 49-50 AD; and his third missionary journey was from 53-57 AD. Obviously he did not spend a great deal of time in one place, it is believed that he spent no more than one year at any one place, at the most three years.

The Apostle Paul had a great deal of faith in the Spirit of God and in the Word of God to bring the Saints to maturity and to expand the cause of Christ to the ends of the world through the Church.

When we think about a Church exploding in growth as a result of our evangelism, the logical practical question would be at what number are we going to look to plant another Church? Remember that the first Church began with 3,000 then 5,000, the book of Acts speaks of the Church in terms of thousands or many to the point of turning the world upside down. There is nothing sacred or spiritual about being a small Church, in fact there is something wrong with a Church where there is no growth. There are what is known as mega-churches. The 100 largest Churches in America in 2013 ranged from 43,500 to 6,010 in attendance.[91] It now depends on ones philosophy of ministry whether to become a mega-church or to stay at a medium size Church. It has been said that it is best to grow to 1,000 in attendance before planting a new Church. Experience has shown that this works well whereas smaller Churches struggle with a Church plant. The rational is simply that with 1,000 it is possible to send out 100 or 10% and still have enough people to sustain the ministry of the base Church. It also means

---

91   Outreach Magazine/LifeWay Research, "The Largest Churches in American 2013," *Outreach,* The Outreach 100, (special issue) 87-88.

that the Church is growing through evangelism, there is an enthusiasm and an energy that has been developed through this growth. Therefore, even though 10% have been sent to plant a new Church, the base Church is growing at least at a growth rate of 10% or more so that in a little more than a year, the base Church will be right back at 1,000. Meanwhile the Church plant, which is made up of people who are in the habit of leading others to Christ will also grow by at least 10% and will now have 110 in attendance. At a mere 10% growth rate the Church plant would be at about 260 in 10 years. If each one at the Church plant led 1 person to Christ per year then after 4 years they would number over 1,000 and would be ready for their own Church plant.

We understand that the Lord is going to carry on His work through His people in His Church (an assembling of His people). It is important to keep in mind that the word *church* as it is used in the New Testament most often refers to the local Church, therefore the emphasis of divinely inspired Scripture is on the local Church.

In Summary:

- Believers come together as a local Church
- The Lord adds to the Church those who would be saved

- The power of the gospel is such that the Church is ever growing
- The Church grows to the point where it plants another one that plants another one etc. to the ends of the world.
- Growth and maturity are dependent upon the Spirit of God and the Word of God
- When by sight there are not enough resources to move forward, faith kicks in and the Lord provides.

Strategy:

It is proposed that our strategy be built upon the following points:

1. Prayer
2. Training
3. Present the gospel
4. Disciple and build up the Saints in the Faith
5. Make the preaching and teaching of the Word of God the priority
6. Equip the Saints for the work of the ministry
7. Move forward by faith

Prayer: It has been said that prayer changes things and this has been proved to be true over the last 2,000 years. The Scriptures have a great deal to say about prayer and praying. For example we see that we are to pray rather than be anxious and the peace of God will guard our hearts (Philippians 4:6-7); we are to pray without ceasing (I Thess. 5:17); we are to pray for all men to be saved (I Timothy 2:1) and we are to pray for one another (James 5:16). In researching various ministries over the years, it is obvious that every ministry which did great things for the Lord spent a great deal of time in prayer.

Prayer is important for at least two reasons. The first is that the Scriptures tell us to pray. Our God is omniscience; therefore we are not informing Him of anything that He does not already know. He is sovereign so He is going to perform His will as He chooses to do so. At the same time we understand that He has chosen to carry out His will through His people and He has told His people to pray. For example our Lord gave an outline for prayer in Luke 11:2-4 and then instructed His disciples and us to ask, seek and knock in verse 9. We are also told in Luke 18:1, ". . . that men always ought to pray and not lose heart". Second, is that this is the way in which our Lord carries out His will in our lives, we pray, He answers, we act accordingly. The way in which we determine the will of God for ourselves and the local Church is through prayer.

Training: The key to this strategy is that each one is able to win one to the Lord and then bring them into the local Church. Obviously some people just cannot bring themselves to give the gospel, therefore we encourage them to at least invite people to come to the local Church where we can give the gospel for them. At the same time we offer training with the goal of equipping the Lord's people in how to give the gospel so that people will know what to say, when to say it, why we say what we say and how to defend what we say. With this training it helps to give confidence to the Lord's people so that they are more apt to give the gospel. This training can be done by having classes as a part of the adult Sunday School and this book can be used as a training manual.

Present the Gospel: The emphasis on presenting the gospel will be on the individual. We begin with each individual being equipped to present the gospel as a normal part of every day life where in the course of conversations, the subject turns toward spiritual things. A part of the training is in how to turn a conversation toward spiritual things and then how to pursue with the gospel. To assist in providing opportunities the Church will offer events that are specifically designed for creating opportunities to present the gospel. Once again, it will be the responsibility of the individual to invite and/or bring someone to the event and then be prepared to give them the gospel. At the same time there will be promotional material and advertising of the event in the community and at each event there will be those who are able to give the gospel. Again, if one does not feel capable of giving the gospel, then all they need to do is to invite one to Church or to an event where one who is comfortable giving the gospel may do so.

Disciple: In Matthew 28:19, disciple is a verb and it is common in this day to speak of disciple as a verb i.e. something that we do. The process of making a disciple begins with the gospel and then moves to a profession of faith and then to fellowship within the local Church. It is in the local Church where the new believer is taught the Scriptures, receives encouragement and care, worships the Lord in the corporate sense, and is built up in the Faith. It is therefore important that we take heed to our curriculum that what needs to be taught to make disciples is actually being taught.

Preach the Word: There was a time in Church history when this would go without saying but that is no longer true. There are many churches that do not preach the Word but rather seek to pacify, amuse or entertain. This strategy goes by a number of different names and will draw in crowds of people who want religion but they do not want it to be convicting. The result is people who have itchy ears and only want to hear what appeases them with the result that they perish for lack of knowledge (Hosea 4:6). In contrast to this, we want to preach the word as is summed up nicely in II Timothy 4:2-4:

Preach the word! Be ready in season and out of season. Convince, rebuke, exhort, with all long-suffering and teaching. For the time will come when they will not endure sound doctrine, but according to their own desires, because they have itching ears, they will heap up for themselves teachers; and they will turn their ears away from the truth, and be turned aside to fables.

The challenge is, will this result in growth that will bring us to over-flowing resulting in a Church plant which will then grow with this same strategy? It is proposed that the answer is yes, that there are more than enough people who want to be taught the Scriptures and to grow thereby. Therefore, this strategy is based upon sound teaching from the pulpit and in our adult Christian education.

Equip the Saints: This is very similar to making disciples but bears special attention in that we not only want to teach people what it means to be a Christian in theory but also in service. We understand that all believers have the Holy Spirit within them and that this Spirit brought spiritual gifts with Him in making us all one body in Christ (I Corinthians 12). Therefore, the next step after learning the Scriptures is to use these Spiritual gifts in service to others. "For even the Son of Man did not come to be served, but to serve, and to give His life a ransom for many." (Mark 10:45). The word *minister* is servant, to *minister* is to serve, a *ministry* is a service (note the word διακονos diakonos). So when we speak of our ministry or of expanding our ministry we are speaking of our service. This service is both inside the local Church and outside into the community. Our teaching and Discipling is to equip the Saints to do this service (Ephesians 4:11-12). Serving within the local Church helps to build up the Saints and to encourage one another. Serving out in the community is the best way to create opportunities for the gospel and to have an irresistible influence on the community to the glory of our Lord. This equipping the Saints for ministry is a key component to this strategy.

By Faith: It is important to understand that the Lord's people do this work by faith as the just shall live by faith (Romans 1:17; Hebrews 10:38). There are going to be times when there is going to be a need for more space, facilities, materials, teachers, pastors, evangelists and money to support all of these. At the same time the demographics studies, the economy and all appearances may indicate that it cannot be done. This has been the experience of nearly every ministry that has ever existed from the time of the resurrection to the present. It is at this moment that those ministries, which have gone on to grow and expand, have stepped out by faith in the living God who has commissioned us all with the task of making disciples of the nations and is capable of providing what is needed to accomplish that task. It has been said that the Lord is simply waiting for His people to move before He steps in and provides what is needed. We see this quite often in the Old Testament where the Lord expected the nation of Israel to step forward by faith before intervening on their behalf. Such examples include when Israel was to enter the land of Canaan the priest has to first step into the waters of Jordan and get their feet wet before the river would part for their crossing (Joshua 3:13). Again, Israel was required to march around Jericho for seven days before the walls came down (Joshua 6:1-20). In the New Testament we see the steps of the Apostle Paul were directed by the Holy Spirit as he was moving forward (see Acts 16:6-10). C.H. Mackintosh wrote concerning the faith of Abraham,

> There is no one so entirely independent of all around as the man who is really walking by faith, and waiting only upon God; but the moment a child of God makes himself a debtor to nature or the world, he loses his dignity, and will speedily be made to feel his loss. It is no easy task to estimate the loss sustained by diverging, in the smallest measure, from the path of faith. No doubt, all those who walk in that path will find trial and exercise; but one thing is certain, that the blessings and joys which peculiarly belong to them are infinitely more than a counterpoise; whereas, when they turn aside, they have to encounter far deeper trial, and naught but that. [92]

---

[92]   C.H. Mackintosh, *Genesis to Deuteronomy Notes on the Pentateuch* (New Jersey: Loizeaux Brothers, 1972) 79.

Then again as we read in Romans 4:20-21, "He did not waver at the promise of God through unbelief, but was strengthened in faith, giving glory to God, and being fully convinced that what He had promised He was also able to perform." Therefore our strategy is one of faith for without faith it is impossible to please God (Hebrews 11:6).

In Summary:

- Individuals give the Gospel and/or invite others to Church
- New people come to Church and are discipled
- Oops—too many people, need more space
- Step out by faith and build more space
- Train a team for the new Church plant
- Reach the point of overflowing
- Step out by faith and send out those equipped to plant a new Church
- Continue to grow to replace those who have been sent out and repeat the process
- Step out by faith and send a team to an unreached people group to plant a Church

Practical Steps:

1. Pray, asking the Lord whether we should build additional space or relocate?

    a. Where does the Lord want this ministry located?

    b. Where is there a neighborhood where we can love, serve and lead the people to Christ?

2. Once we have clear leading from the Lord, by faith, we press forward.

3. Reach a consistent average worship attendance of 600

    a. Hire a staff member in order to train that man to help facilitate growth at the base Church and to become the pastor of the Church plant

    b. Identify property in the desired area for the Church plant

    c. Confidently convey this plan to the base Church membership

    d. This will work concurrently with building more space.

4. Reach a consistent average worship attendance of 800

    a. Hire a worship assistant, training that man to assist with worship at the base Church and to become the worship leader of the anticipated Church plant

    b. Identify a group of approximately 100 from the base Church who will form the nucleus of the new Church plant.

    c. Identify and train the leadership of that new Church plant with the guidance of the then-active Elders, including the pastoral staff.

    d. Those that are a part of the Church plant will need to be those who have demonstrated the ability to bring others into the local Church. This is what distinguishes this strategy from others.

5. Reach a consistent average worship attendance of 1,000

    a. Launch the Church plant

    b. Arrangements made for financial support of the Church plant.

Back to the Beginning:

- We are to make disciples of the nations (people groups) beginning where we are and expanding outward.
- It is the gospel that changes lives who in turn change society
- The Church is God's ordained biblical institution to achieve this purpose, therefore we plant Churches
- Prayer and faith are key to any strategy for growth

- We are to serve one another and the community around us, our service will equip the Saints and lead souls to Christ.

In the practical steps cited above there are some areas which are in need of elaboration. The first has to do with location. It has been said that the three most important aspects of retail marketing are location, location, location. In a sense the Church is in retail marketing and location is important for both the base Church and for the Church plant. It has been said that the most important aspect to the location of a Church is where there is a place where we can love the people, serve the people and lead the people to Christ. Once again it is all about the people. Even the base Church needs to continually evaluate where they are in terms of the community around them. If the community is dying out for one reason or another, will that affect the Church? Could there be a time when it is necessary for the Church to move to a new location where it can be more effective in ministry especially in the ministry of reaching more people for Christ? This is particularly true of a Church plant. The first consideration needs to be where can we serve the people of the community? Consultants have argued that a growing dynamic Church ought to be able to simply transfer its "DNA" to another location and all will be well. Consultants are paid large sums of money to know what they are talking about, therefore to question their wisdom is to tread on thin ice. However, there are mega-churches that have multiple sites and each site is different from the base Church. The reason is that they are reaching a different demographic of people, therefore to serve these people they change their "DNA" to fit how they are going to minister to the people. The base Church may have a large auditorium with cushion chairs lined up in neat little rows. The Church plant might start out in a coffee shop sitting on couches and lounge chairs in a circle. The same message is being preached, the same music is played but the environment is different because the demographics are different. In other words we are seeking to reach the people with the same absolute truth of the Word of God but fitting into the community where we are seeking to love, serve and lead to Christ. Something else to consider is that the base Church may have a long history. This means that there are certain things that the Church will do simply because they have always done it that way. Each new person coming into that Church understands it and accepts it and becomes a part of it. A new Church plant is starting with a clean slate and new people. There is no reason why they cannot develop their own traditions based upon their own experiences and where they are in their maturity in Christ. For example many old Churches are stuck in their support of the traditional method of doing missions. The new Church plant may very well see what God is presently doing to reach the world and get on board with this new wave. So while a Church may plant its "DNA" it may be very well advised to be open to new avenues of ministry that the base Church is unable to explore because of its history and traditions.

Invariably when a Church is growing consistently it will find itself squeezed for space. The need for space is not just because of more people but it can also be because of more ministries that are needed to meet the needs of the more people. For example, an old Church starts to grow because they have captured the vision for personal evangelism. Suddenly they have a need for an Adult Class on the basics of the faith, but there are no classrooms available. It is very common for growth to include young families with babies and small children, suddenly there is a need for more nursery space and Sunday School rooms. Therefore, even before the Church is ready to plant a new Church they have a need for more space, which means a building project and an capital campaign. It may very well be that the elders of the Church, after prayer and due diligence, have come to an agreement that it is the will of God to proceed with a capital campaign. There are some interesting observations that can be made concerning capital campaigns, I will note three of these.

First, it can be easily argued that capital campaigns came out from the world rather than the Church. Particularly in the United States there are nonprofit corporations who do "fund raising" to meet their annual budget. When they have a large project, such as the building of a new building or adding on to an existent one, they will have a "capital campaign." There are actual classes a person can take to learn how to do fund raising and capital campaigns, in fact a person can be licensed as a fund raiser etc. When a Church does a capital campaign they do the exact same things that any nonprofit corporation would do and even use the same language. The only difference is that in the Church there is provision made for some sort of prayer, Scriptures are given as justification and the name of the Lord Jesus is included in the language. A consultant is hired and the cost of the consultant along with the printed materials and mailings etc. adds up to tens of thousands of dollars that could have gone toward the actual project. Now there is certainly nothing innately wrong with this as there are a lot of things the world invented which the Church uses, like electricity etc. The question becomes is there another way for the Church to behave that actually demonstrates faith? Our Lord had told us to ask and it shall be given to us, must this include a campaign? Is it possible for the

Lord's people to ask and the Lord then provides what is needed even for a multi-million dollar project?

This brings me to the second observation and that is the use of Scriptures in a typical campaign. It is certainly important to use Scriptures in the Church particularly on the part of the consultant who needs to biblically justify his existence as a consultant and to biblically justify the whole concept of a campaign. This is where it gets real interesting. Verses in the Bible which teach us about our stewardship and how we are to support the Lord's work are now used to justify a campaign. It is like nobody ever heard of these verses until now that we have a capital campaign, the Holy Spirit has waited until now to reveal to us this startling little fact. Verses that have to do with sacrificial and joyful giving are now being applied to this campaign. The whole concept of faith is wrapped up in the campaign, we clearly have not been living by faith until now and now we have this grand opportunity to exercise faith. I suppose that after the campaign we can go back to our miserly little lives of not giving sacrificially or living by faith. This begs several questions. Are not these verses to be seen as how we are to live our lives every day and not just during a campaign? Have you noticed how that suddenly people have extra money that they can now lay aside for the campaign. According to these very Scriptures, should we not be giving sacrificially every time we come together? It has been said that if every Christian gave just 10% of their income there would be more than enough money to support the Lord's work all over the world. In other words if the Lord's people actually obeyed what the Scriptures teach, there would be plenty of money to build when we need to build and to plant new Churches locally and globally. The fact that a campaign is needed to uncover these funds is not a cause for rejoicing but of despair and the need for repentance. Is it possible for the Holy Spirit to move the Lord's people to give as the Word of God has instructed us to give every week?

The third observation is how *faith* is defined in a campaign. In the corporate world it is not necessary to use the Scriptures, to speak of the Lord Jesus or to use the word *faith*. As previously mentioned fund raising of any kind has been made into a science. There are even mathematical equations which can be used to figure out how to get the money that you need. For example your biggest gift should equal 10 percent of the campaign goal; your first 8 gifts should equal .5%; your next 70 gifts should equal 4.8% and all other gifts should equal 94.7%.[93] In order to get these gifts, you learn how to find the people with large gift potential and then train people to go get those gifts. For the masses, it is a matter of numbers as well because

---

[93] James M. Greenfield, *Fund-Raising Evaluating and Managing the Fund Development Process*, (New York: John Wiley & Sons, Inc., 1991) 179.

you soon learn that if you were to do a mass mailing, for example, then you learn how many pieces need to be sent out to get a certain response to equal a certain amount of money. When I was a salesman I understood that I needed to talk to a certain number of people to get a certain number of interviews which would then equal a certain number of sales and a certain number of dollars of commission for me. To get the income that I needed, it all started with getting in front of that certain number of people and then it went from there. The same principle holds true in fund raising. In the secular world you have to sell people on the integrity of the corporation and the value of the corporation's mission. Then you want to target a niche of people who have an interest in that mission. In the Church you are dealing with people who are already sold on the mission and a part of the sales pitch is that when you give, you are giving by faith in accordance with the teaching of the Scriptures. When people are giving to a campaign in the Church they are obeying Scripture and exercising faith. This is a great incentive, therefore, a campaign will emphasize these elements in campaign marketing materials and individual interviews. So far so good, but what happens when it comes time to count up the pledges and the amount pledged falls short of the goal? In the secular world you go back to your marketing and up your numbers. In the Church, the Lord's people get weary of campaigns fairly quickly. So what can a Church do after all the meetings, committees, dinners and testimonies are done and you have still fallen short of the goal? In other words the dog and pony show is done and you do not have the dollars pledged that you need to accomplish what you have already established as the will of God for this Church? Unfortunately, what is many times done is that the original goal is scaled back to match the pledges. The elders of the Church first came to the Church to announce that this is what is needed to advance this ministry. After spending time in prayer, here is the number that is needed. But oops, we do not have the pledges needed to reach this number, now we have a new number. Apparently our understanding of the will of God was off slightly.

In the Bible *faith* is defined in terms of believing in that which is not seen. In Hebrews 11:1 we read, "Now faith is the substance of things hoped for, the evidence of things not seen." For example we read of Abraham going where he did not know where he was going (Hebrews 11:8). Faith can be defined as believing in that which you cannot otherwise attest to with the five senses. Our salvation is based upon faith where we believe in God who cannot be seen. We also read that the just are to live by faith (Hebrews 10:38) therefore there are many times when we continue to believe in spite of the lack of any physical proof that we have good reason to believe.

We live our lives depending upon the truthfulness of the Word of God which declares the message of an unseen God. Yet when we are confronted with a shortage of funds to do what we believe is the Lord's will, we move back from faith to that which is seen. Is it possible for the Lord's people to trust the Lord to provide what is needed beyond what is pledged?

It is true that the Lord may very well use campaigns to provide what it is needed. After all this is the typical manner in which a Church raises their needed funds or at least a portion of the needed funds. One purpose of this volume is to provide other possibilities and food for thought. Living by faith is a cornerstone principle to the life of the believer so why not place an emphasis on faith and that we believe even when the numbers do not add up. My father used to say that a person has no need to live by faith when he has money in his bank account. A key question for the Church under any circumstances is what place does our Lord, prayer and faith have in what we are doing? Certainly, if we are content with our answer regardless of our method, then all that is left is to wait for the Lord to provide what is needed to carry out His will for the Church.

In the practical steps there are growth markers of 600, 800, and 1,000. These growth markers then trigger certain activity. The steps taken toward a Church plant depend upon growth, reaching others for Christ and training those who will be a part of the Church plant. The ideal formula for Church growth and then Church planting is the Lord's people leading others to Christ or at least bringing them to the Church where the Church can lead them to Christ and then disciple them. At the same time a Church that is active in the community and has a reputation for sound teaching and caring for its members will draw in people from other Churches. This is what is known as transfer growth. There is nothing wrong with this as the Lord's people see where their needs can be met and where they can do ministry more effectively, so they come to this Church. One danger that can arise is that there are those who Church hops i.e. they find something wrong with one Church and then hop to another and then after a while they find something wrong and hop to another etc. The problem is that these kind of people can be disruptive to the momentum of the Church. In these cases, care needs to be exercised and at times it is better to invite these people to go ahead and hop to another Church. In some cases Church discipline may be in order.

Church growth may also be achieved through programming. Nelson Searcy and Kerrick Thomas wrote a book entitled *Launch, Starting a New Church from Scratch*. In this book they cover everything from the first calling from the Lord to start a Church; to vision, goals and a budget; to fund raising; to hiring staff; to beginning with monthly services before going to weekly and on and on. When I say growth achieved through programming it does not exclude evangelism but to the contrary the programming is geared to reaching the lost with a view that the Church will begin with more unbelievers than with believers. For example they write about "launching large":

> Launching large also includes launching from the outside in—which is perhaps the most radical of our "launching successfully" precepts. It is completely possible to launch a church in which the only Christians on the initial team are the staff (pastor, worship leaders and spouses). In building a church from the ground up, you don't have to wait until you can attract a set number of Christians from the area or until you can convince Christians from other areas to embrace your vision and relocate.

> Keeping the goal of launching large in front of you causes a shift in the early DNA of your church— you will have an outward focus from the get-go. Churches that launch large tend to stay focused on the unchurched, while churches that wait to launch often get distracted with insider concerns and the perceived need to "take care of the core." Keeping a new church outwardly focused from the beginning is much easier than trying to refocus an inwardly concerned church.[94]

---

[94] Nelson Searcy and Kerrick Thomas, *Launch Starting a New Church from Scratch*, (Ventura: Regal, 2006) 31-32.

They demonstrate the place that our Lord has in Church planting in the following:

> Launching a new church that impacts the community positively, reaches the lost, grows rapidly, helps people mature in their faith, and then starts more new churches nearby and around the world is entirely possible—with God! When He calls you to start a new church, give all of the potential and possibilities over to Him and let Him lead your work. Then, and only then, will your church become a church of greater success and significance than you ever imagined![95]

You can go to www.churchfromscratch.com to get on their mailing lists for books and aids in starting a Church along with leadership training materials and other helps. Other helpful resources are found at www.Tellstart.com and www.ChurchPlantingSuperSite.com.

One final thought for this chapter. When the base Church plants a Church the Church plant does not need to be sent out on its own. The Apostle Paul thought of the Saints in Thessalonica in terms of parents and children, "But we were gentle among you, just as a nursing mother cherishes her own children." (I Thessalonians 2:7) and ". . . as you know how we exhorted, and comforted, and charged every one of you, as a father does his own children," (I Thessalonians 2:11). The base Church will

---

[95] Ibid., 33.

be as a parent who nourishes the children until such time as they are prepared to leave the nest and move out on their own. But even then there is still a close connection between the parent and the child. In the same manner the base Church will share much of its resources with the Church plant until such time as the plant is ready to move on its own. Even then the Churches will maintain a close bond of fellowship and share in ministry particularly in reaching their local area with the gospel and moving out to the uttermost parts of the world. For example they may be together for youth retreats, short term mission trips, holiday programs, various outreach activities, support for those going to plant another Church and for a team to plant a Church in another country. These Churches will always maintain a bond of fellowship in the gospel.

"Now to Him who is able to do exceedingly abundantly above all that we ask or think, according to the power that works in us, to Him be glory in the church by Christ Jesus to all generations, forever and ever. Amen." (Ephesians 3:20-21

FINAL THOUGHTS

I have always argued that the most significant decade in American history after the turn of the 20th century is that of the 60's. This is the time when a generation challenged the long standing traditions and social norms of this country. This is a time in which the media especially television, went from reflecting culture to creating culture. The news media in particular went from reporting news to creating it or at least the perspective that one should have concerning the events of the day. The greatest of the traditions, norms and values to be discarded was Christianity. In spite of all that the founding fathers did and said to make the United States a country founded upon Christian/biblical principles, the US is no longer a Christian nation and postmodernism is the prevailing philosophy of the day. Postmodernism is more than just a period of time, it is a philosophy based upon humanism, relativism and atheism. In postmodernism the theology is atheism, the philosophy is relativism and the economics is socialism. The way in which postmodernism answers the four great questions of life is as follows:

What is real?	Positivism
What is true?	Relativism
What is the meaning of life?	Skepticism
What happens when we die?	Agnosticism or Epicureanism

To put it simply, postmodernism places mankind at the center of all things, humanity has become its own god. Therefore there is no reality except that which is determined by the individual. There is no truth except that which is believed to be true by the individual at that time, place and for those circumstances which can be changed as time, place and circumstances change. The meaning of life is self determined except that all conclusions must be held in a suspension of doubt. When we die nothing happens, so eat drink and be merry now because there is nothing after this.

So how did our society and culture fall into such a pit of despair? These seeds of change were planted in the 60's and over the course of the following decades the plant has grown and is now bearing its fruit. During the critical time and following the Church took up a castle mentality. The Church withdrew from the realm of public debate and retreated into a philosophical castle where she barred the windows, pulled up the drawbridge and barred the door. No one was welcome into the Church except those who were most like those who were already in it. Should someone, somehow trust Christ and enter the Church, they must conform to those in the castle. This would have been fine if those in the castle were following Christ but instead they were teaching for doctrine the traditions of men. This was not so much different from the relativism outside the castle, truth can be whatever you want it to be. It may be true that the Church of modern times was analytical but it is definitely true that the Church of postmodern times has become ignorant of the Scriptures, the Church is destroyed for lack of knowledge. One of the main reasons that the Church does not engage in public discourse is simply because she no longer knows what to say and even if she does, she is afraid to say it for fear of offending someone. A.W. Tozer wrote of this back in the 70's:

> The church has lost her testimony. She has no longer anything to say to the world. Her once robust shout of assurance has faded away to an apologetic whisper. She who one time went out to declare now goes out to inquire. Her dogmatic declaration has become a respectful suggestion, a word of religious advice, given with the understanding that it is after all only an opinion and not meant to sound bigoted.[96]

Therefore, because of ignorance and fear on the part of the Church the market place is unaware of the Church's message and having only heard the message of

---

[96] A.W.Tozer, *God Tells The Man Who Cares* (Harrisburg PA: Christian Publications, 1970) 35.

postmodernism, our society and culture has embraced postmodernism and all that it entails. The media, specifically television is the loud and persistent voice of postmodernism.

Finally the Church let down the drawbridge and began to enter into the market place. Unfortunately the message of the Church is not the dogma of Scripture that thus says the Lord but rather a compromise with the postmodern world. It can be well argued that the largest Church in America preaches a message of a religious positive mental attitude. As noted in a previous chapter the emergent church lends itself to the relativism of postmodernism that the Bible is merely a story which can be applied however it appeals to the individual. There is no longer sin and repentance but rather just acceptance and compromise. There is no longer the distinction between God loving the sinner but hating the sin. The castle changed its furnishings so that it could look more like the culture, to be culturally relevant. There are books on Church planting and growth which state that one should listen to the radio and see what the culture is listening to and then adapt that to the Church. Some emergent churches do not even bother to adapt it, they just play it as it is.

The premise of *Attack Evangelism* is that it is time to attack culture with the dogmatic, absolute truth of the

Word of God. In the Scriptures the Word is spoken of in terms of a sharp two edged sword which penetrates deep within the soul of man and effects change. The way in which Christians can overcome fear is with the knowledge of the Word and confidence in the power of the Gospel. It is past time for the Church to change culture and to stop being changed by culture.

Now is the time for the disciples of our Lord who take the great commission seriously to think strategically about how we should go on the attack. This should begin with how we think about culture. In the book *Essentials of Sociology* the authors state that culture consists of five basic social institutions:

- The family, to care for dependents and rear children.
- The economy, to produce and distribute goods.
- Government, to provide community coordination and defense.
- Education, to train new generations.
- Religion, to supply answers about the unknown or unknowable.[97]

Obviously the Church does more than provide answers to the unknown or unknowable, in fact the Church provides answers to that which God has determined for

---

[97] David, B. Brinkerhoff and others, *Essentials of Sociology5th ed.* (Belmont CA: Wadsworth Thompson Learning, 2002) 79.

us to know or that which is knowable. The Church goes on to do much more for the individual, the family and the community. The point here is that the Church needs to be involved in ministering (i.e. serving) to families; be involved in economics, in government and in education. These basic institutions of society ought to be our targets for attack. There are primarily two ways in which we can do that. The first is by individual Christians being involved which will lead to the Church being involved. The second is to get into the market place of ideas which is the media. In the New Testament the market place of ideas was in the Synagogues of the Jews and the Agora of the Greek city states. Today philosophy is developed on TV, Radio, Movies and most especially on the Internet. The internet has become the Agora or market place for our world. Therefore, we need to speak up as individuals and as the Church and then we need to become pro-active in the media. This will require brain-storming, and a specific strategy that will grow out of that brain-storming.

Attack Evangelism is being on the offense without being offensive, taking advantage of opportunities to give the gospel as they present themselves in normal living and also creating opportunities where none seemed to exist. Our Lord called upon His disciples to be fishers of men, in order to do any fishing of any kind one must go where the fish are and be proactive in catching them. As our Lord prayed, "As You have sent Me into the world, I also have sent them into the world. And for their sakes I sanctify Myself, that they also may be sanctified by the truth. I do not pray for these alone, but also for those who will believe in Me through their word;" (John 17: 18-20). Therefore, "For I am not ashamed of the gospel of Christ, for it is the power of God to salvation for everyone who believes, for the Jew first and also for the Greek." (Romans 1:16).

BIBLIOGRAPHY

Adams, Weldon, *The Church According To The Scripture* Shreveport: LinWel Publishing, 2000.

Aldrich, Joe, *Life Style Evangelism* Colorado Springs: Multnomah Books, 1993.

Anderson, Jim, *Christian Apologetics A Defense Of The Faith* Springfield MO: Video bible Studies International, 1990.

Bakke, Ray, *A Theology As Big As The City* Downers Grove: Intervarsity Press 1997.

Barton, David, *The Founding Fathers on Jesus, Christianity and the Bible* [article online] (accessed 6 October 2010) available from http://www.wallbuilders.com/LIBissuesArticles.asp?id=8755

Brinkerhoff, David B., and others, *Essentials of Sociology5th ed.* Belmont CA: Wadsworth Thompson Learning, 2002.

Carson, D.A., *Showing The Spirit* Grand Rapids: Baker Book House, 1987.

Christianity Today—General Statistics and Facts of Christianity **http://christianity.about.com/od/denominations/p/christiantoday.htm**

Church or Parachurch [article on line] (accessed 17 December 2010) available from http://triablogue.blogspot. com/2006/01/church-or-parachurch.html; Internet

Eckman, Jim, Issues in Perspective, www. issuesinperspective.com August 8-9, 2009

Evans, C. Stephen, *Philosophy of Religion* Downers Grove: Intervarsity Press, 1982.

Federer, William, *America's God and Country Encyclopedia of Quotations* St. Louis: Amerisearch, INC., 2000.

Ferguson, Dave, "Winning at Any Size," *Outreach* 2010 Special Issue September 2010.

Gibbs, Eddie, and Bolger, Ryan K., *Emerging Churches* Grand Rapids: Baker Academic, 2005.

Geisler, Norman L., *Christian Apologetics* Grand Rapids: Baker Publishing, 1976.

Gitt, Werner, *Counting the* Stars; [article online]; available from http://www.answersingenesis.org/creation/v19/i2/ stars.asp; internet

Got Questions?org, "What is the emerging/emergent church movement?" http://www.gotquestions.org/Printer/ emergent-church-emerging-PF.html [accessed on Oct. 30,2012]

Greenfield, James M., *Fund-Raising Evaluating and Managing the Fund Development Process*, New York: John Wiley & Sons, Inc., 1991.

Haggai Institute, http://www.haggai-institute.com/

Hall, Verna M, ed. *The Christian History of the Constitution of the United States of America* San Francisco: The Foundation for American Christian Education, 1966.

Hendricks, Kevin, *Your Invited: Bringing People to Church* [article online] (accessed 13 October 2010) available from http://www.churchmarketingsucks.com/2005/04/youre-invited-bringing-people-to-church/

Ironside, H. A., *Lectures on the book of Revelation* New Jersey: Loizeaux Brothers, 1975, reprint.

Kelly, William *Lectures on the New Testament Doctrine of The Holy Spirit* Denver : Wilson Foundation, reprint.

Keil, C.F., and Delitzsch, F., *Commentary on the Old Testament*, Vo. 1 Grand Rapids: Eerdmans Publishing, reprint 1975.

Kennedy, D. James, *Why I Believe* Dallas: Word Publishing, 1980.

Kirk, Russell, *The Roots of American Order* Wilmington Delaware: ISI Books, 2003 4th ed.

Koehn, Greg, "The Value of the New Testament Church in a Postmodern World" PhD diss., Louisiana Baptist University, 2013.

Lewis, Robert, *The Church of Irresistible Influence* Grand Rapids: Zondervan Publishing 2001.

MacDonald, William, *Parachurch Organization* [article on line] (accessed 17 December 2010) available from http://web.singnet.com.sg/~syeec/literature/parachurch.html

Mackintosh, C.H., *Genesis to Deuteronomy Notes on the Pentateuch* New Jersey: Loizeaux Brothers, 1972.

McKnight, Scot, *The Blue Parakeet* Grand Rapids: Zondervan, 2008.

Miller, Darrow L, *Discipling Nations* Seattle: YWAM publishers, 1998.

Merriam Webster's Deluxe dictionary, 1994 10th ed., "para."

Morgan, G. Campbell, *The Gospel According to Matthew* New York: Revell Co. 1929.

Morison, Frank, *Who Moved The Stone?* Downers Grove: InterVarsity Press, 1971.

Morris, Henry, M., *The Genesis Record* Grand Rapids: Baker Book House, 1976.

Morris, Thomas V., *Our Idea of God* (Notre Dame: Notre Dame Press, 1991

Moyer, Larry, *Non-threatening Evangelism Training System* Dallas: Evantell Inc, 1992.

Noll, Mark A., *The New Shape of World Christianity*, Downers Grove, IL. IVP Academic, 2009.

Outreach Magazine/LifeWay Research, "The Largest Churches in American 2013," *Outreach,* The Outreach 100, (special issue) 87-88.

Parachurch Organizations, http://en.wikipedia.org/wiki/Parachurch_organization [online]

Pentecost, J. Dwight, *Things To Come* Grand Rapids: Zondervan Publishing, 1958, reprint 1971.

Pink, Arthur W., *Exposition of the Gospel of John* Grand Rapids: Zondervan, 1945.

Ryrie, Charles, C., *The Holy Spirit* Chicago: Moody Press, 1965.

Salter, Darius, *American Evangelism* Grand Rapids: Baker Books 1996.

Searcy, Nelson, and Thomas, Kerrick, *Launch Starting a New Church from Scratch*, Ventura: Regal, 2006.

Sider, Ronald J., *One-sided Christianity?* Grand Rapids: Zondervan Publishing 1993.

Strauch, Alexander, *Biblical Eldership* Littleton: Lewis and Roth Publishers, 1986.

Thayer, Joseph Henry, *Greek-English Lexicon* Grand Rapids: Zondervan, 1974.

Thiessen, Henry C., *Lectures in Systematic Theology* Grand Rapids: Eerdmans Publishing, 1979.

Tozer, A.W., *God Tells The Man Who Cares* Harrisburg PA: Christian Publications, 1970.

Tucker, Robert C., ed., *The Marx-Engels Reader 2nd ed.* New York: W.W. Norton Co. 1978.

Wolston, W.T.P., *The Church: What is it?* Denver: Wilson Foundation, 1982, reprint.

Wuest, Kenneth S., *Wuest's Word Studies Vol.1* Grand Rapids: Eerdmans Publishing Co. 1953.

Yancey, Philip, *The Jesus I Never Knew* Grand Rapids: Zondervan publishing, 1995.